Birds of South Asia
The Ripley Guide
Volume 1: Field Guide

Birds of South Asia
The Ripley Guide
Volume 1: Field Guide

by
Pamela C. Rasmussen and John C. Anderton

Plates by (in alphabetical order)
Jonathan Alderfer, John C. Anderton, Hilary Burn, Albert E. Gilbert, Cynthia House,
Ian Lewington, Larry B. McQueen, Hans Peeters, N. John Schmitt, Thomas Schultz,
Kristin Williams and Bill Zetterström

Plate sponsors:
Mrs. Evelyn Bartlett, Mr. Perry Bass, Mr. Howard Brokaw, Mrs. Jackson Burke,
Mrs. Pamela Copeland, Mr. Wallace Dayton, Drue Heinz Fund, Mr. Paul Mellon,
Mrs. Jefferson Patterson, Mr. Howard Phipps and Mr. David Rockefeller

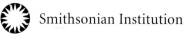

National Museum of Natural History
Smithsonian Institution

in association with

Lynx Edicions

Front cover figures, painted by John C. Anderton. From top to bottom:
  Crimson-backed Flameback *Chrysocolaptes stricklandi*
  Stork-billed Kingfisher, Nicobars race *Pelargopsis capensis intermedia*
  Indian Eagle-owl *Bubo bengalensis*
  Black-and-orange Flycatcher *Ficedula nigrorufa* (male and female)
  Himalayan Quail *Ophrysia superciliosa*

Back cover figures, painted by John C. Anderton:
  Top: Serendib Scops-owl *Otus thilohoffmanni*
  Bottom: Nicobar Scops-owl *Otus alius*.

Recommended citation:
Rasmussen, P.C. & Anderton, J.C. 2005. *Birds of South Asia. The Ripley Guide*. Vols. 1 and 2. Smithsonian Institution and Lynx Edicions, Washington, D.C. and Barcelona.

© 2005. Smithsonian Institution.

All rigths reserved. No part of this book may be reproduced or transmitted in any form or by any means, electronical or mechanical, including photocopy, recording, or any information retrieval system, without the prior written permission of the copyright holders.

Printed and bound in Barcelona by Ingoprint, S.A.
D. L.: B-13.182-05
ISBN Volume 1: 84-87334-65-2
ISBN Volumes 1 & 2: 84-87334-67-9

Contents

List of plates, artists, and plate sponsors 2
How to use the *Field Guide* 9
Field Guide 11
Index to genera and group names in plates and maps (plate index) 372

See **Volume 2: Attributes and Status** for Acknowledgements, detailed maps, text and Appendices.

List of Plates (artist; plate sponsor)

Plate 1	**Divers and grebes** (I. Lewington)	12
Plate 2	**Cape Petrel and shearwaters** (I. Lewington)	14
Plate 3	**Petrels and storm-petrels** (I. Lewington)	16
Plate 4	**Tropicbirds and pelicans** (I. Lewington)	18
Plate 5	**Cormorants and darter** (I. Lewington)	20
Plate 6	**Frigatebirds and boobies** (I. Lewington)	22
Plate 7	**Egrets** (J. Schmitt)	24
Plate 8	**Herons** (J. Schmitt)	26
Plate 9	**Herons and bitterns in flight** (J. Anderton)	28
Plate 10	**Pond- and night-herons** (J. Schmitt)	30
Plate 11	**Striated Heron and bitterns** (J. Schmitt)	32
Plate 12	**Storks** (J. Anderton)	34
Plate 13	**Ibises, spoonbill, storks and flamingos in flight** (J. Anderton)	36
Plate 14	**Ibises, spoonbill, adjutants and flamingos** (J. Anderton)	38
Plate 15	**Swans and geese** (H. Burn)	40
Plate 16	**Whistling-ducks, shelducks and large perching ducks** (H. Burn)	42
Plate 17	**Dabbling ducks** (C. House)	44
Plate 18	**White-headed Duck and bay ducks** (C. House)	46
Plate 19	**Cotton Teal, Marbled Duck, Mandarin Duck and sea ducks** (C. House)	48
Plate 20	**Ducks in flight** (H. Peeters)	50
Plate 21	**Kites and bazas** (J. Schmitt)	52
Plate 22	**Accipiters perched** (J. Schmitt)	54
Plate 23	**Accipiters in flight** (J. Schmitt)	56
Plate 24	***Buteo* buzzards** (J. Schmitt)	58
Plate 25	***Butastur* buzzards, honey-buzzards and serpent-eagles** (J. Schmitt)	60
Plate 26	**Short-toed and *Hieraaetus* eagles** (J. Schmitt)	62
Plate 27	**Hawk-eagles and Black Eagle** (J. Schmitt)	64
Plate 28	***Aquila* eagles perched** (J. Schmitt)	66
Plate 29	***Aquila* eagles in flight** (J. Schmitt)	68
Plate 30	**Osprey, fish-eagles and sea-eagles** (J. Schmitt)	70
Plate 31	***Gyps* vultures** (J. Schmitt)	72
Plate 32	**Other vultures** (J. Schmitt)	74
Plate 33	**Harriers I** (J. Schmitt)	76

Plate 34	**Harriers II** (J. Schmitt)	78
Plate 35	**Falconets, Merlin and kestrels** (J. Schmitt)	80
Plate 36	**Mid-sized falcons** (J. Schmitt)	82
Plate 37	**Large falcons** (J. Schmitt)	84
Plate 38	**Partridges and francolins** (J. Schmitt)	86
Plate 39	**Partridges, hill-partridges, Himalayan Quail and megapode** (J. Schmitt)	88
Plate 40	**Quails and buttonquails** (J. Schmitt)	90
Plate 41	**Spurfowl and junglefowl** (J. Schmitt)	92
Plate 42	**Blood Pheasant, tragopans and monals** (H. Burn)	94
Plate 43	**Shorter-tailed pheasants and snowcocks** (H. Burn)	96
Plate 44	**Long-tailed pheasants and peafowl** (H. Burn)	98
Plate 45	**Cranes** (J. Anderton)	100
Plate 46	**Cranes and bustards in flight** (J. Anderton)	102
Plate 47	**Bustards** (J. Anderton)	104
Plate 48	**Barred and rufous rails** (H. Burn)	106
Plate 49	**Dusky rails, gallinules, coot and finfoot** (H. Burn)	108
Plate 50	**Painted-snipe, recurvirostrids, Crab-plover, oystercatcher and jacanas** (J. Alderfer, J. Schmitt)	110
Plate 51	**Coursers, pratincoles and thick-knees** (J. Alderfer, J. Schmitt)	112
Plate 52	**Lapwings** (J. Alderfer)	114
Plate 53	**Plovers** (J. Alderfer, J. Schmitt)	116
Plate 54	**Curlews, godwits and dowitchers** (J. Alderfer)	118
Plate 55	**Tringine sandpipers** (J. Alderfer)	120
Plate 56	**Turnstone and larger calidrines** (J. Alderfer)	122
Plate 57	**Stints and phalaropes** (J. Alderfer)	124
Plate 58	**Painted-snipe, snipes and woodcock** (J. Schmitt)	126
Plate 59	**Skuas** (I. Lewington)	128
Plate 60	**Large gulls** (I. Lewington)	130
Plate 61	**Small gulls** (T. Schultz)	132
Plate 62	**Pale *Sterna* terns** (T. Schultz)	134
Plate 63	**Large and crested terns** (T. Schultz)	136
Plate 64	**Dark terns and skimmer** (T. Schultz)	138
Plate 65	**Sandgrouse** (H. Burn)	140
Plate 66	**Rock pigeons and imperial-pigeons** (I. Lewington)	142
Plate 67	**Woodpigeons and Nicobar Pigeon** (K. Williams)	144
Plate 68	**Turtle-doves and cuckoo-doves** (I. Lewington)	146
Plate 69	**Emerald Dove and green-pigeons** (K. Williams)	148
Plate 70	**Hanging-parrots and smaller parakeets** (H. Burn)	150
Plate 71	**Ringed and larger parakeets** (H. Burn)	152
Plate 72	**Small cuckoos, crested cuckoos and koel** (J. Anderton)	154
Plate 73	**Typical cuckoos and hawk-cuckoos** (J. Anderton)	156
Plate 74	**Malkohas and coucals** (J. Anderton)	158
Plate 75	**Hawk-owls and barn-owls** (L. McQueen)	160
Plate 76	**Fish-owls, Snowy Owl and eagle-owls** (L. McQueen)	162
Plate 77	***Asio* and wood-owls** (L. McQueen)	164
Plate 78	**Scops-owls** (L. McQueen)	166
Plate 79	**Owlets and Boreal Owl** (L. McQueen)	168
Plate 80	**Frogmouths and nightjars** (J. Anderton)	170
Plate 81	**Nightjars in flight** (J. Anderton)	172
Plate 82	**Treeswift and swifts I** (J. Schmitt)	174
Plate 83	**Swifts II** (J. Schmitt)	176
Plate 84	**Hoopoe, rollers and trogons** (J. Anderton)	178
Plate 85	**Kingfishers I** (A. Gilbert)	180
Plate 86	**Kingfishers II** (A. Gilbert)	182
Plate 87	**Bee-eaters** (J. Anderton)	184

Plate 88	**Hornbills** (A. Gilbert)	186
Plate 89	**Hornbills in flight** (J. Anderton)	188
Plate 90	**Honeyguide and barbets** (J. Anderton)	190
Plate 91	**Wryneck, piculets and small pied woodpeckers** (J. Anderton)	192
Plate 92	**Larger pied woodpeckers** (J. Anderton)	194
Plate 93	**Green and brown woodpeckers** (J. Anderton)	196
Plate 94	**Flamebacks and large woodpeckers** (J. Anderton)	198
Plate 95	**Broadbills and pittas** (J. Anderton)	200
Plate 96	**Bush-, sky- and crested larks** (B. Zetterström)	202
Plate 97	**Finch-, calandra, Horned and hoopoe larks** (B. Zetterström)	204
Plate 98	**Desert larks and short-toed larks** (B. Zetterström)	206
Plate 99	**Martins** (J. Anderton)	208
Plate 100	**Swallows** (J. Anderton)	210
Plate 101	**Wagtails** (J. Anderton)	212
Plate 102	**Olive pipits** (T. Schultz)	214
Plate 103	**Ochre pipits** (T. Schultz)	216
Plate 104	**Woodshrikes, cuckooshrikes and woodswallows** (J. Anderton)	218
Plate 105	**Minivets** (J. Anderton)	220
Plate 106	**Grey and olive bulbuls** (J. Anderton)	222
Plate 107	**Yellow bulbuls** (J. Anderton)	224
Plate 108	**Ioras, leafbirds and fairy-bluebird** (J. Anderton)	226
Plate 109	**Shrikes** (J. Anderton)	228
Plate 110	**Waxwing, Hypocolius, monarchs, fantails and dippers** (J. Anderton)	230
Plate 111	**Cochoas, Grandala, rock- and whistling-thrushes** (T. Schultz)	232
Plate 112	***Turdus* thrushes I** (T. Schultz)	234
Plate 113	***Turdus* thrushes II and blackbirds** (T. Schultz)	236
Plate 114	***Zoothera* thrushes** (T. Schultz)	238
Plate 115	**Shortwings and blue robins I** (J. Anderton)	240
Plate 116	**Bush- and blue robins II** (J. Anderton)	242
Plate 117	**Brown robins, Bluethroat and rubythroats** (J. Anderton)	244
Plate 118	**Shamas, black robin, rock-chat and forktails** (J. Anderton)	246
Plate 119	**Typical redstarts** (L. McQueen)	248
Plate 120	**Large redstarts, water chats, *Irania* and *Hodgsonius*** (L. McQueen)	250
Plate 121	**Wheatears** (L. McQueen)	252
Plate 122	**Bushchats and stonechats** (J. Anderton)	254
Plate 123	**Whistler, brown and pied flycatchers** (J. Anderton)	256
Plate 124	**Grey-backed flycatchers** (J. Anderton)	258
Plate 125	**Blue-and-rufous flycatchers** (J. Anderton)	260
Plate 126	**Blue flycatchers** (J. Anderton)	262
Plate 127	**Laughingthrushes I** (J. Anderton)	264
Plate 128	**Babaxes and laughingthrushes II** (J. Anderton)	266
Plate 129	**Laughingthrushes III** (J. Anderton)	268
Plate 130	**Laughingthrushes IV** (J. Anderton)	270
Plate 131	**Grassland babblers, tit-babbler and *Stachyris*** (J. Anderton)	272
Plate 132	***Turdoides* babblers** (J. Anderton)	274
Plate 133	**Scimitar-babblers** (J. Anderton)	276
Plate 134	**Myzornis, mesias and yuhinas** (J. Anderton)	278
Plate 135	**Fulvettas** (J. Anderton)	280
Plate 136	**Barwings, sibias and white-hooded babbler** (J. Anderton)	282
Plate 137	**Minlas and shrike-babblers** (J. Anderton)	284
Plate 138	**Wren and wren-babblers I** (J. Anderton)	286
Plate 139	**Wren-babblers II and forest-floor babblers** (J. Anderton)	288
Plate 140	**Parrotbills** (J. Anderton)	290
Plate 141	**Cisticolas, *Locustella* warblers and grassbirds** (J. Anderton)	292

Plate 142 **Confusing prinias** (J. Anderton) 294
Plate 143 **Colourful and streaked prinias** (J. Anderton) 296
Plate 144 *Cettia* **bush-warblers** (J. Anderton) 298
Plate 145 *Bradypterus* **bush-warblers** (J. Anderton) 300
Plate 146 **Big or striking reed-warblers** (I. Lewington) 302
Plate 147 **Plain reed-warblers and** *Hippolais* **warblers** (I. Lewington) 304
Plate 148 **Tit-warblers, Goldcrest, tailorbirds and tesias** (J. Anderton) 306
Plate 149 **Bright forest warblers and canary-flycatcher** (J. Anderton) 308
Plate 150 **Brown leaf-warblers** (I. Lewington) 310
Plate 151 **Green leaf-warblers** (I. Lewington) 312
Plate 152 **Leaf-warblers with tertial spots** (I. Lewington) 314
Plate 153 *Sylvia* (I. Lewington) 316
Plate 154 **Grey tits** (J. Anderton) 318
Plate 155 **Long-tailed tits and colourful tits** (J. Anderton) 320
Plate 156 **Creepers and aberrant tit-like birds** (J. Anderton) 322
Plate 157 **Nuthatches** (J. Anderton) 324
Plate 158 **Flowerpeckers** (L. McQueen) 326
Plate 159 **White-eyes and short-tailed sunbirds** (L. McQueen) 328
Plate 160 **Long-tailed sunbirds and spiderhunters** (L. McQueen) 330
Plate 161 **Accentors** (L. McQueen) 332
Plate 162 **Brownish buntings** (L. McQueen) 334
Plate 163 **Yellowish buntings** (L. McQueen) 336
Plate 164 **Mountain-finches and snowfinches** (L. McQueen) 338
Plate 165 **Cardueline finches** (L. McQueen) 340
Plate 166 **Fringillids and desert finches** (L. McQueen) 342
Plate 167 **Rosefinches I** (L. McQueen) 344
Plate 168 **Rosefinches II** (L. McQueen) 346
Plate 169 **Red finches, Gold-naped Finch and bullfinches** (L. McQueen) 348
Plate 170 **Grosbeaks** (L. McQueen) 350
Plate 171 **Avadavats and munias** (L. McQueen) 352
Plate 172 **Sparrows** (L. McQueen) 354
Plate 173 **Weavers** (L. McQueen) 356
Plate 174 **Orioles** (J. Anderton) 358
Plate 175 **Drongos** (J. Anderton) 360
Plate 176 **Starlings I** (J. Anderton) 362
Plate 177 **Starlings II and mynas** (J. Anderton) 364
Plate 178 **Hill-mynas, nutcrackers, jackdaws and choughs** (H. Burn) 366
Plate 179 **Crows and ravens** (H. Burn) 368
Plate 180 **Magpies, jays and treepies** (J. Anderton) 370

How to use the *Field Guide*

This book covers (in two volumes) all species and subspecies of birds (over 2500) known from South Asia, including Bangladesh, Bhutan, India, Maldives, Nepal, Pakistan, and Sri Lanka, as well as Afghanistan and the Chagos Archipelago. For each species, the following elements are provided on the plates and their facing pages in Vol. 1, the *Field Guide*: 1) colour illustrations; 2) identification notes, each with a total-length measurement from a standard source (Ali & Ripley 1983); 3) range maps for each regularly occurring species with annotations on status, usual habitats, and geographic variation if significant. Each species account in Vol. 2 contains far more information on identification, variation, and measurements, and they provide detailed accounts on distribution, habitat, habits, and vocalizations. See Introduction in Vol. 2 for further information on scope, protocols, interpretation of plumage variations, and other aspects. Abbreviations used in labeling figures on the plates are:

Plumage categories:
- br: breeding plumage
- n-br: non-breeding plumage
- 1st-w: first-fall/winter
- 2nd-w: second-fall/winter (approximate)
- juv: juvenile
- ♂: male
- ♀: female
- x: hybrid between indicated species

Regional categories for illustrated races:
- NW: north-western part of region (Afghanistan, Pakistan and/or India)
- N: northern part of region, including Himalayas and northern plains
- NC: north-central India, e.g. Satpura Hills
- SW: south-western India (and often Sri Lanka)
- S: southern India and Sri Lanka
- WG: Western Ghats only
- EG: Eastern Ghats only
- SL: Sri Lanka only
- NE: northeastern India (and often Bangladesh)
- SAH: South Assam Hills, north-eastern India south of the Brahmaputra River
- SE: lowlands of South Assam Hills

Figures without such labeling represent typical adults of the species from a race widespread in the region; color morphs are not labeled as such but this is indicated in the facing notes. All other plumage and regional categories are spelled out on the plates.

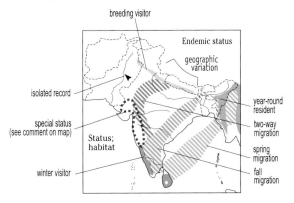

Field Guide

PLATE 1. DIVERS AND GREBES (I. Lewington)

Divers or loons *Gavia*

Large and long-bodied, normally marine birds with short tails, dagger-like bills.

1.1. Black-throated Diver
Gavia arctica   65cm   p.41

Vagrant; Haryana. Rounder-headed than Red-throated with a straighter, thicker bill and more curved neck. Breeding adult has black throat bordered by white stripes, and large white spots on back. Non-breeder is darker above than non-breeding Red-throated, with strongly contrasting head and neck pattern and (often) white flank-patch. In flight, heavier and looks more pied, with feet projecting farther past tail, and bill held more horizontally.

1.2. Red-throated Diver
Gavia stellata   61 cm   p.41

Vagrant; SW Pakistan. Lighter-built than Black-throated, with angular profile and upturned bill. Breeding adult has pale grey head and chestnut throat. Non-breeder is paler above with more white on face and neck, and less contrast than Black-throated. In flight, note short foot projection, sagging neck and hunched back, quick wing-beats.

Grebes *Podiceps*, *Tachybaptus*

Small to medium, with short 'tailless' bodies and thin necks, pointed bills, white wing-patches. Note bill size and shape, dark/light pattern of head.

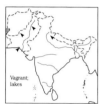

1.3. Horned Grebe
Podiceps auritus

34cm   p.42

In non-breeder, from Black-necked by straighter bill and flatter crown, with pale lores, and dark cap not reaching cheek. See text for breeding adult (unlikely in region). In flight (not shown) has small white forewing-patches and no white on primaries; see Red-necked.

1.6. Red-necked Grebe
Podiceps grisegena

45cm   p.41

Smaller, thicker-necked and darker than Great Crested, with black-tipped yellow bill. Breeding adult has pale cheek and rufous neck; non-breeder has dark cheek and lores. In flight darker above than Great Crested, with less white on forewing.

1.4. Black-necked Grebe
Podiceps nigricollis

33cm   p.43

In non-breeder, from Horned by steep forehead, upturned bill, and darker cheek. Breeding adult has black neck and golden cheek-plumes. In flight white patch on trailing edge reaches inner primaries (longer than in other grebes).

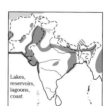

1.7. Great Crested Grebe
Podiceps cristatus

49cm   p.42

Large and pale, with triangular head, and long thin neck and bill. Breeding adult has rufous cheek-ruff. Non-breeder has white around eye, pink bill and white foreneck; in flight much white in secondaries and lesser wing-coverts.

1.5. Little Grebe
Tachybaptus ruficollis

27cm   p.41

Small, brown and short-billed, with black cap and pale rear end. Breeding adult with chestnut cheek and foreneck and yellow gape-wattle. Non-breeder has paler brown face and flanks; juv. has dark stripes on face. In flight, white secondaries.

PLATE 2. CAPE PETREL AND SHEARWATERS (I. Lewington)

Petrel *Daption* others on Plate 3

Chunkier than shearwaters, with shorter, heavier bills, shorter necks and different flight.

2.1. Cape Petrel
Daption capense 39cm p.43

Vagrant; Sri Lanka. Unmistakable: white below and on rump, with black head and tail-tip, and pied wings.

Shearwaters *Puffinus, Calonectris*

Long wings and long thin bills. Note colour of wing-lining, relative length and shape of wings and tail. In maps of these species, yellow indicates non-breeding, mainly summer passage.

Common; atolls, inshore

2.2. Audubon's Shearwater
Puffinus lherminieri
31cm p.46
Small, short-winged and long-tailed with slim mostly dark bill; black above and white below, with white axillaries and wing-lining, black vent and patch on side of breast.

Nearshore

2.3. Persian Shearwater
Puffinus persicus
>31cm p.46
Slightly larger than Audubon's, with longer, broader wings, usually a longer, mostly pale bill, and shorter tail; paler brown above, with less sharp face/throat contrast, usually smaller breast-patches and browner flanks, and darker underwing and axillaries.

Scarce; pelagic

2.4. Streaked Shearwater
Calonectris leucomelas
48cm p.45
Large and broad-winged; dark brown above with pale scaling, white face and underparts, and long pale bill.

2.5. Sooty Shearwater
Puffinus griseus 43cm p.45

Hypothetical; Sri Lanka. Dark with usually conspicuously whitish underwing; see Short-tailed.

2.6. Short-tailed Shearwater
Puffinus tenuirostris 42cm p.45

Vagrant; Pakistan, Sri Lanka. From Sooty by shorter bill with darker wing-lining (but see text), and from dark-morph Trindade Petrel (Pl. 3) by dark base of primaries, different shape and flight.

Pelagic

2.7. Flesh-footed Shearwater
Puffinus carneipes
47cm p.45
Large, all dark, with broad wings, pale bases of flight feathers from below, short broad tail, pink bill and feet. Looks big-headed, full-chested and round-backed on water, wings projecting beyond tail. Heavier than Wedge-tailed, with different shape and flight.

Pelagic

2.8. Wedge-tailed Shearwater
Puffinus pacificus
45cm p.45
Slim with broad-based, rather rounded wings, long narrow tail, and long dark bill. Dark morph is grey-brown, with less striking pale feet and flight-feather bases than Flesh-footed. Light morph (not illustrated; see text) unlikely in region. At rest, looks round-headed with pinched neck; wing-tips do not reach tail-tip.

PLATE 3. PETRELS AND STORM-PETRELS (I. Lewington)

Petrels *Pterodroma, Bulweria, Pseudobulweria*
see also Plate 2

Chunkier than shearwaters with shorter, heavier bills, shorter necks and different flight. None are known to breed in region, and in maps yellow indicates mainly summer passage. See text for additional species hypothetical or possible for Chagos.

3.1. Barau's Petrel
Pterodroma baraui

38cm   p.44

Mid-sized with heavy black bill; pale grey above with black cap, white forehead and face, blackish 'hand' and middle wing-coverts forming a broad M across spread wings, and dark rump and tail. From below, white with grey on breast-sides, black bar and carpal patch through white wing-lining, and black trailing edge to wing.

3.2. Trindade Petrel
Pterodroma arminjoniana   31cm   p.43

Hypothetical; S India. Long-winged with short heavy black bill; pale patch on primary bases recalls skuas. Dark morph otherwise all dark (see Short-tailed Shearwater, Pl. 2); see text for pale morph.

3.3. Jouanin's Petrel
Bulweria fallax

31cm   p.44

Larger than Bulwer's with broader wings and tail, larger bill, paler face; pale wing panels less distinct. Smaller than Wedge-tailed Shearwater (Pl. 2) with narrower wings, stouter, darker-based bill and pinkish feet blackish near tip; see also Mascarene.

3.4. Soft-plumaged Petrel
Pterodroma mollis   35cm   p.44

Hypothetical; Sri Lanka. Grey above and on breast-band with dark eye-patch; wing pattern from above and below recalls Barau's, but note grey crown and pale rump and tail.

3.5. Mascarene Petrel
Pseudobulweria aterrima   p.43

Hypothetical; W. All dark; from smaller Jouanin's by heavier bill, much less wedge-shaped tail, broader wings, and heavier legs and bicoloured feet.

3.6. Bulwer's Petrel
Bulweria bulwerii

27cm   p.44

Small and dark, with long tapered tail and pale bar on secondary coverts; legs pale fleshy, grading into dark toe-tips. See Jouanin's and storm-petrels.

Storm-petrels *Oceanodroma, Oceanites, Fregetta, Pelagodroma*

Small and delicate with short broad wings, and typically fluttering, erratic, low flight, taking food from sea surface. Some follow ships. Note especially wing pattern, underparts and rump colour and pattern, and foot projection; also flight style. None are known to breed in region, and in maps yellow indicates mainly summer passage.

3.7. Swinhoe's Storm-petrel
Oceanodroma monorhis

20cm   p.47

Long, angled wings with weak pale secondary-covert bar; feet do not project beyond moderately forked tail. Pale primary-shafts much less marked than on Matsudaira's. Fast flight.

3.8. Matsudaira's Storm-petrel
Oceanodroma matsudairae

24cm   p.47

Hypothetical; occurs in SW waters. Large with pale bar on secondary coverts, white patch at base of primary shafts, and deeply forked tail. Slow wing-beats.

3.9. Wilson's Storm-petrel
Oceanites oceanicus

17cm   p.46

Shortish rounded wings, pale upperwing-band, white rump and rear flanks (less extensive than on Black-bellied), and squared tail; feet project beyond tail.

3.10. Black-bellied Storm-petrel
Fregetta tropica

20cm   p.46

Curved white rump-band, squared tail, weak pale bar on secondary coverts, (normally) broadly white rear flanks and wing-linings, black vent and dark feet that trail behind. From White-bellied by black belly-stripe (hard to see); see text.

3.11. White-bellied Storm-petrel
Fregetta grallaria   20cm   p.47

Hypothetical; occurs in SW waters. From Black-bellied by lack of black belly-stripe, (usually) more white in wing-lining, and less foot projection.

3.12. White-faced Storm-petrel
Pelagodroma marina

20cm   p.46

Grey above with dark eye-patch on white face, dark breast-sides, narrow pale bar on secondary coverts, broad pale grey rump, short broad black tail, and long leg extension; white below and on wing-lining.

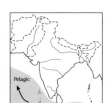

PLATE 4. TROPICBIRDS AND PELICANS (I. Lewington)

Tropicbirds *Phaethon*

Mainly white, with dagger-like bills and long tail-streamers. Note pattern of upperparts.

4.1. White-tailed Tropicbird
Phaethon lepturus

39cm (w/o streamers)   p.47

Small, with black outer primaries but white primary coverts, and yellow bill. Adult has black wing-bar. Juv. has widely spaced fine barring above, a dusky yellow bill and no nape-band.

4.2. Red-billed Tropicbird
Phaethon aethereus

48cm (w/o streamers)   p.48

Barred above (looking silvery at a distance), with black outer primaries and primary coverts; heavier-billed and broader-winged than White-tailed. Adult has orange bill; juv. has yellow bill, fine barring above.

4.3. Red-tailed Tropicbird
Phaethon rubricauda

39cm (w/o streamers)   p.48

Broad wings with little black. Adult has red streamers and bill. Juv. has coarse, widely spaced barring above and no nape-band.

Pelicans *Pelecanus*

Huge waterbird with very long pouched bills, long necks, short legs, long broad wings, and short tails. Note facial feathering and colours, crests.

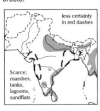

4.4. Spot-billed Pelican
Pelecanus philippensis

152cm   p.49

Relatively small; from Dalmatian by less feathered face, darker eye, dark spots on pale bill/pouch, and scruffier head and neck. Breeding adult has dark face, pinkish rump and vent. In flight as Dalmatian but secondaries from below darker, tail brown.

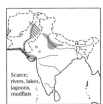

4.5. Dalmatian Pelican
Pelecanus crispus

183cm   p.49

Huge; greyish with feathered face and pale iris and eye-ring. Breeding adult has long crest, bright pouch, black legs. Juv./imm. browner above. In flight, pale secondaries and tail.

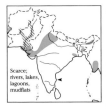

4.6. Great White Pelican
Pelecanus onocrotalus

183cm   p.49

Size of Dalmatian, with extensive pinkish to orange facial skin, yellow-orange pouch and dark eye, pale feet. Breeding adult has drooping crest. Juv. dark above with yellow pouch. In flight, remiges dark; juv. from below also has dark leading edge of wing.

Plate sponsored by Mrs. Jefferson Patterson.

PLATE 5. CORMORANTS AND DARTER (I. Lewington)

Cormorants *Phalacrocorax*

Large to small, dark waterbirds with long bodies and necks with small heads and hooked bills. Note facial feathering and colours, bill size and shape, feather patterns above, and relative length of tail. On maps for this family, reported breeding areas indicated in green, other occurrences in blue.

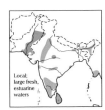

Local; large fresh, estuarine waters

5.1. Indian Shag
Phalacrocorax fuscicollis

63cm   p.52

Mid-sized and slender, with low crown and long thin bill; more facial skin than Little Cormorant but less than Great Cormorant, and no pale lore-spot. Breeding adult has white cheek-tuft; non-breeding has white throat. Juv. has indistinct black scales above, and some dark on belly (usually white in juv. Great). In flight slender with mid-length tail.

Large fresh marshes

5.3. Pygmy Cormorant
Phalacrocorax pygmeus

51cm   p.51

Often not separable from Little. Breeding adult is brown-headed with blacker, less silvery tone to upperparts; non-breeding often with larger, less well-defined white throat-patch than Little. Juv. has less distinctly white throat than juv. Little, and whitish or white-streaked belly.

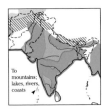

To mountains; lakes, rivers, coasts

5.2. Great Cormorant
Phalacrocorax carbo

80cm   p.52

Large, thick-necked and big-headed, with heavy hooked bill, extensive orange facial skin, and broad white chin-strap up to cheek. Breeding adult has much white on head; non-breeding has diffuse face-patch. Juv. is dark brown with (usually) solid white belly. In flight, head and neck large, tail relatively short.

Common; fresh wetlands, villages

5.4. Little Cormorant
Phalacrocorax niger

51cm   p.52

Very like Pygmy, which see. From Indian Shag by short bill thicker at base, steeper forehead and pale lores. Breeding adult black, silvery above; non-breeding browner, with white throat. Juv. usually dark below, with pale fringes. In flight note long tail.

Darter *Anhinga*

Cormorant-like but with stiletto-like bill on small head, very long thin neck, long tail and broad wings.

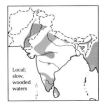

Local; slow, wooded waters

5.5. Oriental Darter
Anhinga melanogaster

90cm   p.53

Extremely long thin neck and bill, boldly white-striped upperparts and black underparts. Adult has dark crown and white neck-stripe, male with black-spotted throat. Imm. and juv. duller. In flight, long crooked neck, broad wings and tail.

PLATE 6. FRIGATEBIRDS AND BOOBIES (I. Lewington)

Frigatebirds *Fregata*

Relatively small-bodied seabirds with extremely long pointed wings and forked tails, small feet and long hooked bills. Note light/dark pattern of underparts and underside of wing; immatures may be inseparable. On maps for this family, yellow indicates regular non-breeding occurrence in summer.

Boobies *Sula, Papasula*

Large and long-bodied seabirds with long and narrow, pointed wings, pointed tails and longish but stout bills. Note overall light/dark pattern (especially from below), and soft-part colours.

6.1. Lesser Frigatebird
Fregata ariel

76cm   p.54

Smaller than congeners and lightly built, with white axillaries. Male has white running to side of belly. Female has white hindcollar and breast between black throat and black oval on belly; lacks female Christmas's black 'spur' on upper side of breast. Juv. has pale hood (rusty on head) to white belly.

6.2. Christmas Frigatebird
Fregata andrewsi

95cm   p.54

Much larger than Lesser, with long heavy bill. Male has white mid-belly-patch but black axillaries. Female similar to female Lesser but with white lower belly, black 'spur' before white axillaries, and black of throat running further onto breast. Juv. like juv. Lesser, but has more white on axillaries forming spur.

6.3. Great Frigatebird
Fregata minor

93cm   p.54

Large with black axillaries. Male black with paler wing-coverts above and reddish feet. Female has diagnostic pale throat and no hindcollar, straight border between white breast and black belly. Juv. has pale head, and more white on belly than juv. Lesser.

6.4. Red-footed Booby
Sula sula

41cm   p.51

Small with pale face. Adult white to pale brown, but with white rump, vent and tail, and red feet; scapulars are pale and black of flight feathers does not reach body. White morph has black carpal patch from below; brown morph and juv. have dark underwing. Juv. has dark face and bill, brown tail and hindquarters, sometimes with paler vent.

6.5. Brown Booby
Sula leucogaster

76cm   p.50

Dark brown with white belly and pale bill and feet; pale wing-lining with dark leading edge. Juv. pale brown below rather than white; lacks pale hindcollar of juv. Masked.

6.6. Masked Booby
Sula dactylatra

80cm   p.50

Large with heavy pale bill and dark face. Adult is white with black flight feathers (reaching body) and tail; from below, lacks black carpal spot of white Red-footed. Juv. from juv. Brown by white hindcollar and upper breast; has white wing-lining with dark stripe but thinner dark leading edge.

6.7. Abbott's Booby
Papasula abbotti   p.51

Hypothetical; Chagos. Large and white with black eye-patch, scapulars, tail and thigh-patch, and narrow wings entirely dark from above and mostly white from below with black-tipped primaries. Juv. (not shown) has black replaced by dark brown.

Plate sponsored by Mr. Perry Bass.

PLATE 7. EGRETS (J. Schmitt)

See Plate 9 for flight illustrations

Egrets *Egretta, Bubulcus*

White or polymorphic, mostly smaller herons with complex seasonal and age variations in plumes and soft part colours. Note overall shape and proportions: especially leg and bill length (and colour), shape of neck.

Estuaries, mangroves, reefs

7.1. Western Reef-heron
Egretta gularis

63cm　　　　　　　　p.55

Slim and gracile, with long neck and long, partly pale bill; may be white, grey or patchy. White morph told from smaller Little by heavier yellower bill and legs; in breeding, finer dorsal plumes. Non-breeding lacks plumes; dark morph is slaty-blue with large white throat-patch, and blacker legs. Juv. has duller bill and more olive legs; dark-morph juv. is paler, patchier grey than adult.

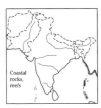

Coastal rocks, reefs

7.2. Pacific Reef-heron
Egretta sacra

58cm　　　　　　　　p.55

Oddly shaped: heavy-billed and long-necked with short thick legs; two morphs. Adult white morph has mostly yellow bill and legs; adult dark morph is dark slaty with inconspicuous white throat-stripe. Juv. pale morph is flecked with dark; juv. dark morph brownish, sometimes with pale scales above.

Common; wetlands

7.3. Little Egret
Egretta garzetta

63cm　　　　　　　　p.55

Small and delicate with long slim black bill, black legs and yellow feet; smaller than Intermediate, with longer bill and flatter forehead. Breeding adult has long narrow nape-plumes and filmy dorsal plumes. Non-breeding adult has pale lower edge of lower mandible; juv. similar, with yellowish rear stripe on legs, and duller feet.

Common; inland wetlands

7.4. Great Egret
Egretta alba

90cm　　　　　　　　p.56

Large; white with long heavy bill and dark gape-line, flat crown, sharply kinked neck, and heavy knees. Breeding adult has black bill, reddish legs, and filmy plumes on back only. Non-breeding adult and juv. lack plumes and have yellow bill, blackish legs. Larger in NW.

Wetlands

7.5. Intermediate Egret
Egretta intermedia

80cm　　　　　　　　p.56

Mid-sized, but not much smaller than some female Great; rounder head with higher forehead, shortish bill, and gape-line to mid-eye; softly kinked neck, and dark legs and feet. Slimmer-necked and lankier than Eastern Cattle. Breeding adult has black bill, and long filmy chest- and back-plumes. Non-breeding adult and juv. have yellow bill with small black tip, and lack plumes.

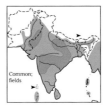

Common; fields

7.6. Eastern Cattle Egret
Bubulcus coromandus

51cm　　　　　　　　p.58

Stocky and thick-necked, with large head and shortish pale bill; legs rather thick and often blackish. See Intermediate. Breeding adult has bright buffy head, neck and back, and orange bill; see Squacco Heron (Pl. 10). Non-breeder white with yellow bill (no black tip) and dark legs; juv. has duskier or pinker bill. In Chagos, see text for introduced Western *B. ibis* (p.58), which is similar but stockier, in breeding plumage with orange-buff mainly on crown, breast and mantle.

PLATE 8. HERONS (J. Schmitt)

see Plate 9 for flight illustrations

Herons *Ardea*

Larger species, mostly grey, some accented with rufous; no colour morphs, and seasonal variation less marked than in egrets, limited largely to head plumes and facial skin colouring. Note especially colour and pattern of neck.

Common; wetlands

8.1. Grey Heron
Ardea cinerea

96cm   p.56

Rather large and mostly pale grey with white head, black crown-stripe and foreneck-stripes; no rufous in plumage. Juv. duskier but has ghost of adult patttern.

8.4. Great-billed Heron
Ardea sumatrana   110cm   p.57

Hypothetical; Nicobars. Large with deep blackish bill (looking slightly upturned); rather plain dark grey, including belly. Adult is all grey, including belly. Juv. more rufescent, with pale streaks below.

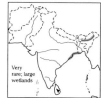

Very rare; large wetlands

8.5. Goliath Heron
Ardea goliath

147cm   p.57

Massive, with very heavy dark bill and legs; grey with chestnut head and neck lacking plumes and side-stripes. Juv. variably chestnut above, with streaked belly and brownish thighs.

Local; reedbeds, swamps

8.2. Purple Heron
Ardea purpurea

86cm   p.57

Mid-sized and very lanky with long thin rufous neck and long bill, shortish yellow legs. Adult has black crown and neck-stripes lacking in much larger Goliath. Juv. rufescent overall with reduced neck-stripes and scaly upperparts.

Near-endemic

Rare; forested streams, rivers, marshes

8.3. White-bellied Heron
Ardea insignis

127cm   p.57

Very large and long-necked with dark bill and legs; dark grey with white throat, belly and vent, and white streaks on neck and breast. Juv. browner with pale fringes above.

PLATE 9. HERONS AND BITTERNS IN FLIGHT (J. Anderton)

See Plates 7, 8, 10 and 11 for standing illustrations; see text for juveniles

9.1. Purple Heron
Ardea purpurea p.57, Pl. 8

Folded neck forms large square bulge, and large feet obvious. No white leading edge to wing, which has little contrast above and below (somewhat more in juv.); dark rufous wing-lining less conspicuous than on Goliath.

9.2. Goliath Heron
Ardea goliath p.57, Pl. 8

Immense, with broad wings lacking contrast above, below with extensive dark rufous wing-lining; heavy legs project far.

9.3. White-bellied Heron
Ardea insignis p.57, Pl. 8

Dark grey with paler rump, white belly and wing-linings with dark remiges.

9.4. Great-billed Heron
Ardea sumatrana p.57, Pl. 8

Uniquely all grey.

9.5. Grey Heron
Ardea cinerea p.56, Pl. 8

Folded neck is less square and bulging than in Purple, feet less heavy. From above, pale wing-coverts contrast with dark flight feathers; from below, wings dark grey, with striking white leading edge and carpal-spot.

9.6. Great Egret
Egretta alba p.56, Pl. 7

White with large neck bulge, broad wings, and long heavy dark legs.

9.7. Intermediate Egret
Egretta intermedia p.56, Pl. 7

White with broad rounded wings; smallish neck bulge, short bill and long dark legs.

9.8. Pacific Reef-heron
Egretta sacra p.55, Pl. 7

Very broad rounded wings and short legs (with large yellow feet) give unique shape.

9.9. Western Reef-heron
Egretta gularis p.55, Pl. 7

Wings more pointed, not as broad as in Pacific, with longer leg projection. White morph like Little, but bill is pale.

9.10. Little Egret
Egretta garzetta p.55, Pl. 7

White, with moderate leg projection and yellow feet. Very like white morph Western Reef but with slender dark bill.

9.11. Eastern Cattle Egret
Bubulcus coromandus p.58, Pl. 7

Smaller neck bulge than Intermediate or Little; narrower and more pointed wings than Intermediate, and from Little by shorter pale bill. In breeding, from pond-herons by narrower wings with white wing-tips and upper mantle, and greater leg extension.

9.12. Squacco Heron
Ardeola ralloides p.58, Pl. 10

Wings slightly more pointed than in pond-herons, with less body/wing contrast.

9.13. Indian Pond-heron
Ardeola grayii p.58, Pl. 10

Broad white wings and dark upperparts, and only feet project (*vs* legs in Eastern Cattle). Juv. has duskier wing-tips, as in other juv. *Ardeola*.

9.14. Chinese Pond-heron
Ardeola bacchus p.59, Pl. 10

Wings slightly longer, more pointed than in Indian; otherwise inseparable except in breeding plumage, when has black mantle.

9.15. Black-crowned Night-heron
Nycticorax nycticorax p.59, Pl. 10

Chunky with broad rounded wings and short legs. Adult is pale with black cap and mantle; juv. is dark with white trailing edge to wing.

9.16. Striated Heron
Butorides striata p.59, Pl. 11

Uniformly dark with longish pointed wings.

9.17. Eurasian Bittern
Botaurus stellaris p.62, Pl. 11

Large and mostly pale with long wings, from above with pale coverts and barred flight feathers.

9.18. Chestnut Bittern
Ixobrychus cinnamomeus p.61, Pl. 11

Wings very broad, round-tipped and entirely chestnut; considerable leg extension.

9.19. Yellow Bittern
Ixobrychus sinensis p.60, Pl. 11

Like Little but with pale mantle, and black of flight feathers does not meet body; feet project further, and wings slightly more rounded.

9.20. Little Bittern
Ixobrychus minutus p.60, Pl. 11

Dark above with pale fore-wing panel; wings rather pointed, and only toes trail tail. From below, pale wing-lining and dark remiges (as in Yellow).

9.21. Malayan Night-heron
Gorsachius melanolophus p.60, Pl. 10

Wings broad with cinnamon forewing, black flight feathers and diagnostic pale trailing edge and tip. Note also black crown and pale mantle, and black-and-white spots on primary coverts.

9.22. Black Bittern
Dupetor flavicollis p.61, Pl. 11

Dark above with little or no contrast in broad, rounded wings.

PLATE 10. POND- AND NIGHT-HERONS (J. Schmitt)

See Plate 9 for flight illustrations

Pond-herons *Ardeola*

Medium-sized herons with relatively short legs and heavy bills and thick necks; primarily white with brown on neck and upperparts, with plainer non-breeding plumages that can be inseparable. Note especially degree of streaking on neck, weight and length of bill and legs.

Common; any wetland

10.1. Indian Pond-heron
Ardeola grayii

46cm p.58

Thickset with rather short stout bill, yellow on lower mandible, and dark line on lores. Breeding adult buff with maroon mantle and blue facial skin. Non-breeder and juv. have heavily striped head and neck, and brown mantle with pale streaks on scapulars; essentially identical to non-breeding Chinese.

Status unclear

Scarce; wetlands

10.2. Chinese Pond-heron
Ardeola bacchus

52cm p.59

Not normally distinguishable from Indian except in breeding plumage; slightly larger, longer-billed, with more bare facial skin and larger eye-ring, and no dark line on lores. Breeding adult has chestnut head and breast, and black mantle.

Marshes

10.3. Squacco Heron
Ardeola ralloides

46cm p.58

Hypothetical; Afghanistan. More lightly built than Indian and Chinese Pond, with longer bill and shorter tarsi. Breeding adult is apricot with blue face and bill-base, and striped nape-plumes lacking in Eastern Cattle Egret (Pl. 7). Non-breeding has browner mantle (paler than in non-breeding pond-herons), and finely streaked head and neck with no plumes. Juv. like juv. pond-herons, but paler above and more finely streaked on foreparts.

Night-herons *Gorsachius, Nycticorax*

Stocky, mid-sized, crepuscular herons, both distinctive.

Scarce; forested streams, swamps

small

10.4. Malayan Night-heron
Gorsachius melanolophus

51cm p.60

Mid-sized, chunky, with shortish dark bill and lax crest; unstreaked in all plumages. Adult is cinnamon with black crest. Juv. is grey, finely barred and vermiculated, with white-tipped crest.

Common; wooded wetlands, villages

10.5. Black-crowned Night-heron
Nycticorax nycticorax

58cm p.59

Stocky, with heavy bill and legs, large red or yellow eye. Adult black above with grey wings, and white below. Juv. brown and heavily streaked on head and underparts; from all pond-herons by white spots on mantle.

PLATE 11. STRIATED HERON AND BITTERNS (J. Schmitt)

See Plate 9 for flight illustrations

Striated Heron *Butorides*

Small, distinctive, somewhat bittern-like greyish heron, extremely variable in region.

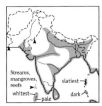

11.1. Striated Heron
Butorides striata

44cm   p.59

Small with yellowish legs and dark bill; grey with black cap and pale edges on greenish mantle. Varies from whitish in Maldives through medium-grey on mainland, to blackish in Andamans. Juv. browner and heavily streaked below, speckled above; from juv. bitterns by more solid upperparts and greyer wing.

Bitterns *Dupetor, Ixobrychus, Botaurus*

Rather large to very small primarily brown herons with short legs; streaked on neck in female, subadult and non-breeding plumages, but never white below as in pond-herons. Note bill shape and pattern, neck pattern.

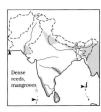

11.2. Black Bittern
Dupetor flavicollis

58cm   p.62

Mid-sized; very dark with long bill and pale neck-patch. Male black above with yellow neck-sides; female rich brown. Juv. paler, with pale-edged wing-coverts.

11.5. Little Bittern
Ixobrychus minutus

36cm   p.60

Mid-length bill, dark cap and upperparts with pale wing-panel. Male has black crown and back, buff panel; female browner with pale edgings above (*vs* streaks in female Yellow), chestnut carpal. Juv. pale-fringed above (not streaked), and often heavily streaked below.

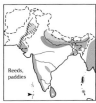

11.3. Chestnut Bittern
Ixobrychus cinnamomeus

38cm   p.61

Stouter than Little and Yellow, with short heavy bill and no wing contrast. Male is chestnut with white submoustachial stripe, red face and yellow bill; female is browner, speckled above and streaked below, with black stripe on cutting edge of bill. Juv. female is very dark, buff-marked above and streaked below.

11.6. Eurasian Bittern
Botaurus stellaris

71cm   p.62

Large and tawny with black cap and moustache, and shortish yellow bill.

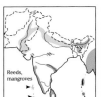

11.4. Yellow Bittern
Ixobrychus sinensis

38cm   p.60

From Little by longer bill, sides of head buffier (*vs* greyish), and less marked wing-panel. Male has dark grey crown and brown back; female is browner with pale streaks above (not just pale edges). Juv. is more heavily streaked above than juv. Little, and usually less heavily streaked below.

PLATE 12. STORKS (J. Anderton)

See Plate 13 for flight illustrations, Plate 14 for adjutants

Storks *Anastomus, Ciconia, Mycteria, Ephippiorhynchus*

Large wading birds with relatively heavier bodies and shorter necks than most herons; neck (except adjutants) held straight out in flight. More graceful than cranes with longer bills. Note bill shape, wing pattern, overall pattern of light and dark.

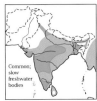
Common; slow freshwater bodies

12.1. Asian Openbill
Anastomus oscitans
76cm   p.63
Pale grey, with black flight feathers and scapulars, thick pale bill with gap in center, short reddish legs. Juv. is greyer with brown mantle, smaller dark bill.

Slow fresh waters

12.5. Painted Stork
Mycteria leucocephala
93cm   p.62
Tall, slim, with long yellowish bill, bare red head, black-and-white wing-coverts and belly-band. Juv. darker and browner with pale bill.

12.2. Oriental Stork
Ciconia boyciana   110cm   p.63
Hypothetical; NE. From White by longer black bill, white eye with red eyeline, and silver trim on secondaries.

Marshes, wooded wetlands

12.6. Black Stork
Ciconia nigra
98cm   p.63
Black with white belly, red bill, eye-patch, and legs. Juv. duller with darker soft parts.

rare in S

12.3. White Stork
Ciconia ciconia
105cm   p.63
White with black wings and scapulars, and red bill and legs. Juv. has blackish bill.

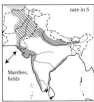
Marshes, fields

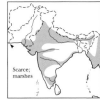
Scarce; marshes

12.7. Black-necked Stork
Ephippiorhynchus asiaticus
135cm   p.63
Extremely tall, with deep long black bill, black head, neck, wing-coverts and lower back. Juv. has brown neck and back.

12.4. Woolly-necked Stork
Ciconia episcopus
83cm   p.63
Black with fuzzy white neck and lower underparts. All ages similar.

Scarce; wooded swamps, streams

PLATE 13. IBISES, SPOONBILL, STORKS AND FLAMINGOS IN FLIGHT (J. Anderton)

See Plates 12 and 14 for non-flight illustrations

13.1. Indian Black Ibis
Pseudibis papillosa p.65, Pl. 14

Bulky with broad wings and tail and little leg extension, all dark except white shoulder from above.

13.2. Glossy Ibis
Plegadis falcinellus p.65, Pl. 14

All dark and spindly, with pointed wings.

13.3. Black-headed Ibis
Threskiornis melanocephalus p.65, Pl. 14

White with dark head and legs; juv. has narrow blackish wing-tips (like juv. spoonbill).

13.4. Eurasian Spoonbill
Platalea leucorodia p.66, Pl. 14

Holds neck outstretched. Adult white with longish dark bill; juv. has dark wing-tips.

13.5. Painted Stork
Mycteria leucocephala p.62, Pl. 12

Broad wings are mostly black; note long neck and leg extension. Below with pale bars to wing-lining and dark belly-band, above with white bar on greater secondary coverts and black tail. Duller in juv., with dark neck.

13.6. Asian Openbill
Anastomus oscitans p.63, Pl. 12

Foreparts white or grey with pale wing-coverts, black remiges and tail; heavy bill and short legs.

13.7. White Stork
Ciconia ciconia p.63, Pl. 12

White with black flight feather and alula; wing-tip squared, with less leg, neck and bill extension than Oriental, and more than Asian Openbill.

13.8. Oriental Stork
Ciconia boyciana p.63, Pl. 12

Foreparts white; lanky, with silvery wing-patch from above; wing-tips perhaps rounder than White's, with more leg, neck and bill extension.

13.9. Black Stork
Ciconia nigra p.63, Pl. 12

Black with white underparts and axillaries.

13.10. Lesser Adjutant
Leptoptilos javanicus p.64, Pl. 14

All dark from above, small white axillary spur from below, with whitish undertail-coverts; folded bare neck can mimic pouch of Greater.

13.11. Greater Adjutant
Leptoptilos dubius p.64, Pl. 14

Pale greater coverts show from above, with white ruff covering upper mantle; from below like Lesser, but with grey undertail-coverts.

13.12. Woolly-necked Stork
Ciconia episcopus p.63, Pl. 12

In flight, alternating dark/light pattern, with moderate leg extension; tail looks white with black edges.

13.13. Black-necked Stork
Ephippiorhynchus asiaticus p.63, Pl. 12

White with long dark neck and legs and broad black stripe on wing-coverts. Juv. (not shown) has darker remiges, but still paler than dark band across coverts.

13.14. Lesser Flamingo
Phoeniconaias minor p.66, Pl. 14

Like Greater but carmine areas more limited on both wing surfaces; black remiges nearly meet body. Juv. best told by shape and bill colour.

13.15. Greater Flamingo
Phoenicopterus roseus p.66, Pl. 14

More gangly than Lesser, with more carmine on wing-coverts, more black in underwing; black remiges do not nearly meet body.

PLATE 14. IBISES, SPOONBILL, ADJUTANTS AND FLAMINGOS (J. Anderton)

See Plate 13 for flight illustrations

Ibises and spoonbills Threskiornithidae

Mid-sized wading birds with rather short legs and necks, and graceful shape; less lanky than storks and cranes, with necks held straight in flight. Note overall colour, bill shape.

Fresh and salt marshes

14.1. Glossy Ibis
Plegadis falcinellus

60cm   p.65
Slender and dark, with thin-based bill, yellowish legs. Breeding adult coppery with green-glossed wings; non-breeder and juv. duller with white streaks on head and neck.

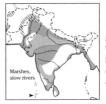

Marshes, slow rivers

14.3. Eurasian Spoonbill
Platalea leucorodia

82cm   p.66
White with black spoon-shaped bill, black legs. Breeding adult has full nape-crest, saffron breast, and orange throat. Juv. has paler soft parts.

Fresh marshes, fields

14.2. Black-headed Ibis
Threskiornis melanocephalus

69cm   p.65
White with naked black head and bill, grey dorsal plumes. Juv. has feathered head.

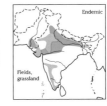

Endemic

Fields, grassland

14.4. Indian Black Ibis
Pseudibis papillosa

73cm   p.65
Heavy-set and dark, with thick bill, naked head with red nape, white shoulder-patch, shortish legs. Juv. has dark head and unglossed upperparts.

Adjutants *Leptoptilos*

Very large grotesque-looking storks, both with naked head and neck, heavy pale bill, grey upperparts and white underparts. To separate, note bill shape, crown pattern, presence of neck ruff and wing panel.

Scarce; fields, swamps, mangroves

14.5. Lesser Adjutant
Leptoptilos javanicus

110cm   p.64
More gracile than Greater, with slimmer bill pale at base, pale skullcap, and blacker upperparts with white trim on coverts but no wing-panel. Juv. duller, brownish above.

recent records within dashes

Rare; wetlands, tips, villages

14.6. Greater Adjutant
Leptoptilos dubius

130cm   p.64
Huge, with swollen dark-based bill, unspotted white neck-ruff, grey wing-panel on grey upperparts. Adult has orange neck-skin and pouch, bare red patch at back of neck. Juv. browner, with more feathered head, and grey-trimmed wing-coverts.

Flamingos Phoenicopteridae

Extremely spindly, tall wading birds, pink as adults and brown as immatures; bill strongly 'bent'. To separate, note especially bill and wing patterns.

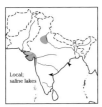

Local; saline lakes

14.7. Lesser Flamingo
Phoeniconaias minor

90cm   p.66
Shorter neck and legs than Greater with an all-dark bill, face and eye; adult brighter pink, with darker red legs. Juv. greyish-white, with black bill and face.

Erratic; saline flats, jheels, estuaries

14.8. Greater Flamingo
Phoenicopterus roseus

110cm   p.66
Extremely tall; pinkish-white to nearly white with pale base of bill, pale iris. Juv. is greyish with browner head and neck, brown streaks on upperparts, and grey-based bill.

PLATE 15. SWANS AND GEESE (H. Burn)

Swans *Cygnus*

Very large long-necked waterfowl, all pale (white as adults). Note neck shape, bill shape and colour, and in flight, foot extension. Not to scale with geese (4–9).

15.1. Bewick's Swan
Cygnus bewickii
122cm   p.68
Small with rather short straight neck, rounded head and concave bill. Adult has more than half-black bill with small yellow patch. First-winter evenly grey, with much pale pink on bill. In flight, shortish neck and tail.

15.3. Mute Swan
Cygnus olor
152cm   p.68
Large with curved neck and dished pale bill with black base and lores. Adult has black knob on orange bill. First-winter unevenly greyish with no knob on grey bill and pale border around facial skin. In flight toes do not project beyond tail.

15.2. Whooper Swan
Cygnus cygnus
152cm   p.68
Long straight neck, triangular head and undished mostly pale bill with black tip. Adult has large yellow bill-patch. First-winter evenly grey. In flight toes project beyond tail.

Geese *Branta, Anser*

Rather large, long-necked, heavy-bodied, mostly grey waterfowl. Note head pattern and neck colour. Not to scale with swans (1–3).

15.4. Red-breasted Goose
Branta ruficollis   61cm   p.71

Hypothetical; C, NE. Small with short thick neck and stubby bill; black with chestnut cheek and breast. In flight looks dark with white vent and flank-stripe.

15.7. Bean Goose
Anser fabalis   76cm   p.69

Winter vagrant; NC. Brownish with dark head and neck and paler breast, dark bill with orange subterminal band, and dark eye-ring. Adult has heavy white edges above, less so in first-winter. In flight no pale wing-patches or dark belly marks; very dark lower back.

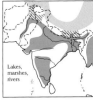

15.5. Bar-headed Goose
Anser indicus
75cm   p.70
Pale grey with white face and neck-stripe, and dark crown-bars, neck and rear flanks. In flight, pale with dark flight feathers and flank-patch.

15.8. Greater White-fronted Goose
Anser albifrons
68cm   p.70
Medium-sized; brownish-grey with a rather dark head and neck, pinkish bill, orange legs and no eye-ring. Adult has white forehead not pointed on crown (unlike Lesser) and irregular black belly marks, lacking in first-winter. In flight dark wings and lower back.

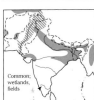

15.6. Greylag Goose
Anser anser
81cm   p.70
Large and pale grey with head and neck little darker than breast, and pink bill, legs and eye-ring; adult has small dark marks on whitish belly. In flight pale wings with dark flight feathers and pale lower back and tail, recalling Bar-headed. See text (p. 71) for hypothetical Snow *Chen caerulescens*.

15.9. Lesser White-fronted Goose
Anser erythropus
53cm   p.70
Small and short-necked with stubby bill and wing-tips extending beyond tail; dark with yellow eye-ring. Adult has white forehead pointed on forecrown and dark belly marks; first-winter lacks both. In flight from Greater by shorter neck, relatively longer wings, quicker wing-beats.

Plate sponsored by Mr. Perry Bass.

PLATE 16. WHISTLING-DUCKS, SHELDUCKS AND LARGE PERCHING DUCKS (H. Burn)

Whistling-ducks *Dendrocygna*

Rather small, brownish, long-necked ducklike birds. Note especially head, neck, and wing patterns.

Scarce; lowland wetlands
vagrant W of red dashes

16.1. Fulvous Whistling-duck
Dendrocygna bicolor

51cm   p.67
Larger than Lesser, paler crown and darker hindneck, striped flank-plumes and whitish neck-patch. In flight, dark wings and white rump.

Common; lowland wetlands

16.2. Lesser Whistling-duck
Dendrocygna javanica

42cm   p.68
Smaller than rarer Fulvous, browner with darker crown and plainer flanks. In flight, chestnut forewing and rump.

Shelducks *Tadorna*

Rather large, strikingly marked but dissimilar-looking ducklike birds.

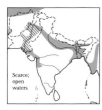

Scarce; open waters

16.3. Common Shelduck
Tadorna tadorna

61cm   p.71
White with black head and wings, red bill and chestnut breast-band. Male has red-knobbed bill; knob lacking in female. In eclipse both approach juv., with dusky crown, hindneck and patches above. In flight white with black head and flight feathers.

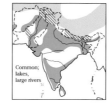

Common; lakes, large rivers

16.4. Ruddy Shelduck
Tadorna ferruginea

66cm   p.71
Orange-chestnut with whitish head and dark rump and tail. Male has black collar. In flight, white wing-coverts above and below, with black flight feathers and tail.

Perching ducks *Sarkidiornis, Asarcornis*

Miscellaneous waterfowl, the two here large and with pied plumage. Small 'perching ducks' on Plate 19.

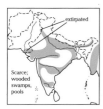

Scarce; wooded swamps, pools
extirpated

16.5. Comb Duck
Sarkidiornis melanotos

76cm   p.72
Large, white with black upperparts and freckles on head and neck, and dark bill and legs. Male has large bill-comb, smaller in non-breeding and lacking in smaller female. Juv. duller with dark eyeline, mottled underparts. In flight, pale with solid dark wings.

Rare; secluded forested wetlands

16.6. White-winged Duck
Asarcornis scutulata

81cm   p.71
Large and very dark, with freckled white head, large white and pale blue wing-patches. In flight dark, with white coverts (above and below) and blue secondaries (above) on broad wings.

Plate sponsored by Mrs. Evelyn Bartlett.

PLATE 17. DABBLING DUCKS (C. House)

See Plate 20 for flight illustrations

Dabbling ducks *Anas*

Small to mid-sized, mostly freshwater waterfowl, sexes often dissimilar, males with eclipse plumage after breeding.

17.1. Common Teal
Anas crecca

38cm   p.76
Compact and small-billed. Male grey with rectangular dark head, lengthwise black-and-white side-stripe, and black-edged buffy rear-patch. Female plain with white at side of tail-base, dark crown and eyeline, no marked pale patches on face. First-winter plainer.

17.2. Andaman Teal
Anas albogularis

43cm   p.75
Small and round-headed, dark brown with pale scales, bluish bill; usually prominent white on face. Juv. has at least whitish throat and eye-ring.

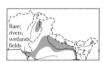

17.3. Baikal Teal
Anas formosa

40cm   p.76
Male has yellowish face and cheek on dark head, white vertical side-stripe, long striped scapulars, black rear end. Female has white spot at bill-base with dark border (unlike female Garganey), broken buffy brow, dark eye-stripe above buffy cheek-stripe; dark flank-spots have pale central marks. First-winter plainer-faced.

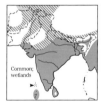

17.4. Garganey
Anas querquedula

41cm   p.75
Male greyish with white brow on purplish head, striped scapulars and brownish, spotted breast and rear; for eclipse see text. Female has dark crown and eye-stripe, pale spot at bill-base, and white throat; first-winter has drabber face.

17.5. Falcated Duck
Anas falcata

51cm   p.73
Large puffy head, longish dark bill, and stumpy rear. Male pale grey with dark head, white throat, black collar, long striped tertials, black rear. Female rufescent with fine buffy scales and plain grey-brown head.

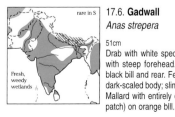

17.6. Gadwall
Anas strepera

51cm   p.73
Drab with white speculum, pale head with steep forehead. Male grey with black bill and rear. Female buffier with dark-scaled body; slimmer than female Mallard with entirely dark culmen (not patch) on orange bill.

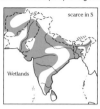

17.7. Eurasian Wigeon
Anas penelope

49cm   p.74
Rather plain head, breast and sides, small pale grey bill. Male is pale grey with white wing-patch, chestnut head, yellow forehead and black rear end; see text for mostly russet eclipse. Female unpatterned with plain pinkish-buff or greyish flanks, whitish belly.

17.8. Northern Shoveler
Anas clypeata

51cm   p.75
Very long heavy bill. Male white with dark head and rear, large chestnut flank-patch. Female resembles female Mallard but for long bill, plainer head.

17.9. Northern Pintail
Anas acuta

56–74cm   p.75
Long slim neck, pointed tail and pale grey bill. Male pale with dark brown head, white neck, and long striped scapulars. Female is buffy with heavy dark scales on body and plain head.

17.10. Chinese Spot-billed Duck
Anas zonorhyncha   61cm   p.74

Vagrant; NE. Less scaled than Indian with double face-stripes, no red bill-spot.

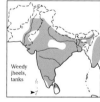

17.11. Indian Spot-billed Duck
Anas poecilorhyncha

61cm   p.74
Dark with heavy pale scales, white wing-patch, pale head with dark cap and eye-stripe, black bill with yellow tip and red basal spot.

17.12. Mallard
Anas platyrhynchos

61cm   p.74
Male is pale grey with dark head, breast and rear, yellow bill and white neck-ring. Female is buffy, heavily streaked darker, with dark eye-stripe; orange bill has black patch on upper mandible.

PLATE 18. WHITE-HEADED DUCK AND BAY DUCKS (C. House)

See Plate 20 for flight illustrations

Stiff-tailed duck Oxyura

Small, stocky, large-billed ducks with longish, stiff tails often held cocked.

Scarce; saline lakes

18.1. White-headed Duck
Oxyura leucocephala

44cm   p.68

Compact and dark with swollen bill, pale face with dark crown and long tail. Breeding male chestnut with white face and blue bill. Female has dark face-stripe, greyish bill.

Bay ducks Aythya, Netta, Rhodonessa

Mid-sized, mostly dark-plumaged diving ducks, winter visitors to open water. Inclusion of *Rhodonessa* in group equivocal.

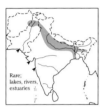

Rare; lakes, rivers, estuaries

18.2. Greater Scaup
Aythya marila

46cm   p.79

Like Tufted Duck but male has paler back, female a much larger white frontal face-patch, paler mid-body; imm. drabber than female with reduced white on face.

Open, deep waters

18.3. Tufted Duck
Aythya fuligula

43cm   p.79

Smaller-billed than Greater Scaup. Male black with long nape-tuft, white sides. Female dark brown with vague paler frontal face-patch, short tuft; imm. plainer with tiny tuft.

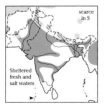

Scarce; lakes, big rivers

18.4. Baer's Pochard
Aythya baeri

46cm   p.79

Dark, with white at front of scaly flanks and white vent; bigger-billed than Ferruginous Duck. Male has dark green head, white iris, pale-tipped grey bill. Female darker-headed than female Ferruginous, with paler patch at bill-base.

scarce in S

Sheltered fresh and salt waters

18.5. Ferruginous Duck
Aythya nyroca

41cm   p.79

Smallish and dark with white vent. Male chestnut with white iris. Female duller with dark iris.

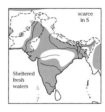

scarce in S

Sheltered fresh waters

18.6. Common Pochard
Aythya ferina

48cm   p.78

Peaked crown, sloping forehead and long grey bill. Male is white with black breast and rear, dark chestnut head. Female grey with darker breast and rear, pale eyeline, face and throat; first-winter more uniformly dark, lacking face pattern.

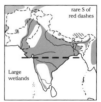

rare S of red dashes

Large wetlands

18.7. Red-crested Pochard
Netta rufina

54cm   p.78

Round head, brown back and longish bill. Male has red bill, orange-rufous head, black breast and rear. Female has dark cap, pale cheek, pale-tipped black bill and uniform brown body.

Near-endemic extinct?

Pools in tall grass jungle

18.8. Pink-headed Duck
Rhodonessa caryophyllacea

60cm   p.78

Dark with long pale neck and bill. Male blackish with pink neck; female browner with dark hindneck.

PLATE 19. COTTON TEAL, MARBLED DUCK, MANDARIN DUCK AND SEA DUCKS (C. House)

See Plate 20 for flight illustrations

Perching ducks and **Marbled Teal** *Nettapus, Marmaronetta, Aix*

Very diverse, distinctive small ducks; large perching ducks on Plate 16.

Common; weedy lakes, ponds

19.1. Cotton Teal
Nettapus coromandelianus

33cm   p.73

Tiny and pale, with small bill, dark cap and upperparts. Breeding male white with black cap, upperparts and breast-stripe, dark vent; eclipse male and juv. duller. Female browner with dark eyeline.

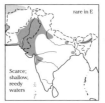

rare in E

Scarce; shallow, reedy waters

19.2. Marbled Duck
Marmaronetta angustirostris

48cm   p.78

Pale with round head, dark bill and eye-patch, buff-spotted body, pointed whitish tail.

Vagrant; wooded pools

19.3. Mandarin Duck
Aix galericulata

48cm   p.73

Lax crest, white eye-stripe. Male has chestnut cheek-ruff and wing-fans, dark breast and golden sides. Female brownish with white spectacles, lores, throat and flank spots.

Sea ducks *Bucephala, Clangula, Mergellus, Melanitta, Mergus*

Small to mid-sized diving ducks, males mostly pied, females browner. Mostly rare winter visitors to coasts, large rivers. Includes mergansers (*Mergellus, Mergus*), fishing ducks with narrow, serrated bills.

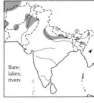

Rare; lakes, rivers

19.4. Common Goldeneye
Bucephala clangula

46cm   p.80

Triangular head with pale eye, short bill, white patch on folded wing. Male white with dark head, large white spot below eye, black back and rear end. Female grey with solid brown head, yellow-tipped black bill.

Vagrant; open water

19.5. Long-tailed Duck
Clangula hyemalis

42cm (w/o tail-pin)   p.80

Short triangular bill, breast darker than sides. Winter male white with dark cheek-patch, black breast, long black tail. Winter female has pale face and sides, dark cheek-patch. First-winter has plainer cheek, browner body.

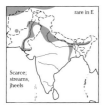

rare in E

Scarce; streams, jheels

19.6. Smew
Mergellus albellus

46cm   p.80

Small with bushy square head and small bill. Male white with black eye-patch, nape and back. Female grey with chestnut head and white throat.

19.7. White-winged Scoter
Melanitta fusca   55cm   p.80

Hypothetical winter vagrant; NW. Dark with sloping forehead; white wing-patch. Male black with white eye-patch, black knob on yellow bill. Female brown with two pale face-patches. In flight dark with white secondaries, heavy body.

Rare; large water bodies

19.8. Red-breasted Merganser
Mergus serrator

58cm   p.81

Thin-based bill. Male has white collar, rusty breast, black on sides of breast; female has double-pointed crest, dull rufous head grading into whitish throat and neck.

Clear fresh water

19.9. Common Merganser
Mergus merganser

66cm   p.81

Deep-based bill extends onto forehead. Male white with black head and back; female has single-pointed crest, rufous head sharply marked from white throat and grey neck. First-winter duller, and sometimes more similar to female Red-breasted.

PLATE 20. DUCKS IN FLIGHT (H. Peeters)

See Plates 17–19 for non-flight illustrations

20.1. Cotton Teal *Nettapus coromandelianus*   p.73, Pl. 19
Tiny, with short rounded wings; male has mainly white remiges and black coverts. Female has dark wings with white trailing edge to secondaries and inner primaries.

20.2. Marbled Duck *Marmaronetta angustirostris*   p.78, Pl. 19
Whitish secondaries, pale wings and tail.

20.3. Mandarin Duck *Aix galericulata*   p.73, Pl. 19
Chunky, long-tailed, large-headed; wings dark with pale trailing edge to secondaries, white belly.

20.4. Northern Pintail *Anas acuta*   p.75, Pl. 17
Long neck and tail, with brownish underwing. Female (shown) uniform with double white border on speculum; male has chestnut upper border to speculum, and white underparts with dark head and rear.

20.5. Mallard *Anas platyrhynchos*   p.74, Pl. 17
Blue speculum with double white border; male has dark–light–dark pattern, female (shown) has brown belly.

20.6. Indian Spot-billed Duck
Anas poecilorhyncha   p.74, Pl. 17
From above wing like Mallard's but speculum green and much white in tertial; from below strong contrast between bright white wing-lining and blackish remiges.

20.7. Chinese Spot-billed Duck
Anas zonorhyncha   p.74, Pl. 17
Vagrant. Like Indian but speculum purple; looks darker-bodied.

20.8. Gadwall *Anas strepera*   p.73, Pl. 17
From above, diagnostic white patch on secondaries; from below, white wing-linings and belly contrast with grey sides.

20.9. Eurasian Wigeon *Anas penelope*   p.74, Pl. 17
Looks round-headed, short-necked and long-winged; from below largely white with pale underwing. Male from above has large white wing-patch above large dark speculum; female has drab upperwing with weak pale upper border to speculum.

20.10. Northern Shoveler *Anas clypeata*   p.75, Pl. 17
Looks very front-heavy, with blue (male) or grey (female) forewing, broad white bar on greater coverts, and green speculum without white trailing edge; underwing mostly white. Male has dark–light–dark–light pattern; female drab with white tail-edge.

20.11. Falcated Duck *Anas falcata*   p.73, Pl. 17
Chunky and short-tailed with brown belly; wings pale grey above, large dark speculum with broad leading white border and small rear one; mostly pale below lacking dark leading edge. Male has dark head and rear end, female is uniform.

20.12. Baikal Teal *Anas formosa*   p.76, Pl. 17
From above, dark speculum with broad white trailing edge; from below, light–dark underwing pattern better marked than in Common, less so than Garganey, with white belly. Male has rather long pointed tail and black vent.

20.13. Garganey *Anas querquedula*   p.75, Pl. 17
From above, pale blue (male) or grey (female) forewing, narrow speculum with broad double white border; from below, wing strikingly black and white, with white belly.

20.14. Andaman Teal *Anas albogularis*   p.75, Pl. 17
From above dark speculum with broad white leading edge; from below, underwing dusky with whitish axillary spur.

20.15. Common Teal *Anas crecca*   p.76, Pl. 17
From above dark speculum with narrow white rear and broad pale upper and rather dark forewing; from below, most of wing-linings white with dark leading edge, and white belly.

20.16. Pink-headed Duck
Rhodonessa caryophyllacea   p.78, Pl. 18
Heavy and dark with long neck, short tail and wide wings, from above with pale secondaries and white leading edge; from below, pale pink underwing.

20.17. Red-crested Pochard *Netta rufina*   p.78, Pl. 18
From above, long bright white wing-stripe on broad wings; from below, wings fairly pale. Male has white leading wing-edge from above, and black central belly stripe from below; in female white axillaries contrast with dark underparts.

20.18. Common Pochard *Aythya ferina*   p.78, Pl. 18
Wings lack contrast, with weak wing-stripe; looks mainly pale with darker foreparts and rear, less so in female (shown).

20.19. Greater Scaup *Aythya marila*   p.79, Pl. 18
Long white wing-stripe nearly reaching wing-tip contrasts with dark forewing; palest at mid-body (female shown). See Tufted Duck.

20.20. Tufted Duck *Aythya fuligula*   p.79, Pl. 18
Very like Greater Scaup except for darker forewing and back; female shown.

20.21. Long-tailed Duck *Clangula hyemalis*   p.80, P19
Compact and variable, but largely pale with dark wings (immature shown). In flight shape compact, with all-dark wings, largely pale body.

20.22. Baer's Pochard *Aythya baeri*   p.79, Pl. 18
Long wing-stripe (weak on outer primaries) and undertail-coverts, underparts and sides white. Male shown; female and young have less contrast below.

20.23. Ferruginous Duck *Aythya nyroca*   p.79, Pl. 18
Longer white wing-stripe than on other bay ducks, and white underwing, belly and vent, otherwise uniform.

20.24. Common Goldeneye *Bucephala clangula*   p.80, Pl. 19
From above, white secondaries and middle coverts on dark wing; from below black underwing and white (male) or grey (female, shown) body.

20.25. White-headed Duck *Oxyura leucocephala* p.68, Pl. 18
Body all dark; wings dark above, with white lining below.

20.26. Common Merganser *Mergus merganser*   p.81, Pl. 19
Long body. Male looks white with dark head and wing-tip. Female (shown) is grey with white secondaries and greater coverts, more than on female Red-breasted.

20.27. Red-breasted Merganser *Mergus serrator* p.81, Pl. 19
Above, white middle and greater secondaries (female, shown) separated by black bar; male has more white on secondary coverts, dark breast and upperparts.

20.28. Smew *Mergellus albellus*   p.80, Pl. 19
Male white with black remiges. Female (shown) grey with pied head and wings.

PLATE 21. KITES AND BAZAS (J. Schmitt)

Kite *Elanus*

Small, whitish, delicate-looking; can recall a gull or male harrier.

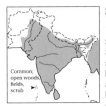

Common; open woods, fields, scrub

21.1. Black-winged Kite
Elanus caeruleus

33cm   p.83

Small; white with long wings and black shoulder. Juv. has rufous wash on breast. In flight wings pointed, from above silvery with black shoulder, from below white with black primaries; tail shortish and white. See male Pallid Harrier (Pl. 33).

Bazas *Aviceda*

Rather small, crested, broad-winged forest raptors with mild expressions.

Scarce; forest

21.2. Jerdon's Baza
Aviceda jerdoni

48cm   p.82

Medium-sized; wing-tips reach more than halfway to tail-tip. Rather plain-headed with long white-tipped black crest, strong mesial streak and broad chestnut bands below. Juv. has dark streaks on head and breast, more spotted underparts. In flight very broad, rounded wings, whitish below with cinnamon wing-lining and barred primary-tips; tail has two dark bands and broad subterminal in adult, four equal-width bands in juv. See short-crested Crested Goshawk (Pl. 22) and much larger hawk-eagles (Pl. 27).

Scarce in S; forest, edge

21.3. Black Baza
Aviceda leuphotes

33cm   p.82

Small and black, with long crest, and white belly (barred chestnut), scapular spots and wing-patch. Female has less white in wing and more chestnut bars below than male. In flight wings paddle-shaped, and black from below with pale grey primaries; longish tail is black above, pale grey below.

Kite *Haliastur*

Rather large, evenly proportioned raptor of cities and water. Adult distinctive but juvenile recalls several unrelated species.

Common; near water

21.4. Brahminy Kite
Haliastur indus

48cm   p.86

Medium-sized. Adult unmistakable: chestnut with white head and breast and black wing-tips. Juv. is pale to dark brown with whitish head and pale-streaked underparts, pale edges above, and plain brown tail. In flight, juv. has chestnut wing-linings, dark secondaries and white primaries with black tips; from above, some chestnut on inner primaries.

Kites *Milvus*

Slim, long-winged and long-tailed, highly aerial raptors.

21.5. Red Kite
Milvus milvus   61cm   p.84

Hypothetical; widely sight-recorded. Rufous with a pale head; similar in shape to Black, but adult has deeper tail-fork and more extensively yellow bill. Juv. (not shown) has shallow tail-fork, dark iris and bill, but is paler-headed than Black, without dark cheek-patch. In flight underwing has large, square white panel in primaries (in some Black), and wing-lining rufous at leading edge, blackish behind; from above, tail is rufous.

Abundant; around humans

21.6. Black Kite (part)
Milvus migrans migrans/govinda

61cm   p.84

Large, dark brown with long, moderately forked, lightly barred tail. Adult is fairly uniform brown, with yellow iris and cere and black bill. Juv. is dark-eyed with pale edges above and pale streaks below; tail less forked. In flight, lanky shape, with long wings usually held flexed; underwing has crescent-shaped pale primary panel (often stronger in juv.), and tail can appear square when spread.

Open areas, wetlands

21.7. 'Black-eared' Kite
Milvus [*migrans*] *lineatus*

Larger and broader-winged than Black, with dark eyes at all ages, pale crown and throat with dark cheek, and bolder dark streaks below; often also a strong rufous suffusion, pale lower belly and vent, more extensive pale wing-patches, and shallower tail-fork. Juv. is more heavily streaked pale below than juv. Black, with bluish cere.

PLATE 22. ACCIPITERS PERCHED (J. Schmitt)

See Plate 23 for flight illustrations

Accipiters *Accipiter*

Rather long-tailed, short-winged and small-billed raptors, mostly of forested habitats in which they prey largely on birds. Most are grey above as adults, often with rufous below, and brown and heavily streaked as juveniles; see hawk-cuckoos (Pl. 73), which mimic both these plumages. Note colour of upperparts, pattern of underparts (especially throat) and facial pattern.

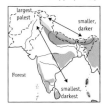

22.1. Besra Sparrowhawk
Accipiter virgatus

32cm   p.99

Small with short primary projection; dark above (grey in male, brown in female) with heavy mesial streak, streaked or solid rufous breast and rufous-barred belly. Juv. has heavy teardrops on breast and thighs, spots on belly and bars on flanks; sometimes rufous on belly and vent. In Andamans and Nicobars male is solid rufous-grey below with grey lower flanks and thighs, white wing-lining and lightly barred flight feathers.

22.2. Shikra
Accipiter badius (part)

35cm   p.97

Rather small with moderate primary projection. Pale grey with at least a hint of a mesial stripe, rufescent bars on breast and belly, and whitish thighs and vent; female has heavier barring below. Juv. is brown above with rufescent scales and heavily streaked crown and cheek, with lighter streaking and barring below than juv. Besra Sparrowhawk.

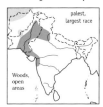

22.3. *Accipiter badius cenchroides*

Mostly migrant NW *cenchroides* (not illustrated) largest and palest; smallest and much darker in Sri Lankan nominate.

22.4. Japanese Sparrowhawk
Accipiter gularis

29cm   p.98

Small and dark, with medium primary projection; absent or weak gular stripe. Male is slaty above, nearly unmarked pale rufous below. Female is dark brown above with heavy grey-brown bars below; lacks pale brow of female Eurasian. Juv. is brown above with black-and-rufous streaks and bars below, finer and paler than in Besra.

22.5. Chinese Sparrowhawk
Accipiter soloensis

30cm   p.98

Small and pale, with very long primary projection; blue-grey above with swollen orange cere, orange legs and unmarked vinous-rufous breast (very dark to dark). See male Andaman race of Besra. Juv. is dark brown above with fine rufous scales, black cap, mesial streak and heavy streaks on breast, bars on flanks and thighs.

22.6. Nicobar Sparrowhawk
Accipiter butleri

30cm   p.97

From Shikra by little or no mesial streak and purer white lower underparts. Juv. uniquely bright chestnut, with dark streaks on throat, chestnut streaks on breast, spots on belly and thighs, and three prominent black bands on tail. Juv. Chinese has darker crown, longer wings, and conspicuous yellow cere.

22.7. Northern Goshawk
Accipiter gentilis

50–61cm   p.99

Large, with fairly long primary projection; dark grey with blackish crown and cheek, prominent white brow wide behind eye (narrower behind eye in female Eurasian Sparrowhawk) and finely barred grey breast. Female is browner above. Juv. is dark brown, mottled pale especially on greater wing-coverts, with short brow (may be lacking), and buffy and completely streaked below, including thighs.

22.8. Eurasian Sparrowhawk
Accipiter nisus melaschistos

34cm   p.99

22.9. *Accipiter nisus nisosimilis*

Smallish to mid-sized (female), with moderate primary projection, no mesial streak, barred underparts. Male is dark blue-grey above with rufous bars below, sometimes solid rufous breast. Female has white brow, narrow behind eye (see Northern Goshawk), and is barred below, usually some brownish on cheek. Juv. browner. Resident Himalayan *melaschistos* is longer-tailed than migrant *nisosimilis*; male is more rufescent below, and both sexes (especially female) are usually darker above.

22.10. Crested Goshawk
Accipiter trivirgatus

43cm   p.97

Chunky with short primary projection and heavy legs and feet; nuchal crest not prominent. Note blackish cap, white throat, strong moustache and mesial streaks, streaked breast and barred lower breast, belly and thighs; female has heavier streaks on breast. Juv. has dark streaks on cheek, and relatively sparse brown streaks on breast and belly; lacks pale areas in greater coverts of juv. Northern.

54

PLATE 23. ACCIPITERS IN FLIGHT (J. Schmitt)

See Plate 22 for perched illustrations

23.1. Besra Sparrowhawk
Accipiter virgatus p. 99

Short wings with blunt rounded tip, strongly barred wing-lining and heavily banded flight feathers, and relatively short, square-tipped tail with broader dark bands than in other accipiters, conspicuous from above.

23.2. Shikra
Accipiter badius p.97

Wings somewhat tapered with narrow rounded tip, and only lightly banded from below; tail is relatively long. Male has unmarked whitish wing-lining and blackish wing-tips (Chinese Sparrowhawk has more pointed wings with extensive black tips) and faint tail-banding from above, none on central rectrices; female has rufescent, lightly marked underwing-coverts, and tail often has a dark subterminal band and 1–3 dark spots on central feathers; juv. has very lightly spotted ochraceous wing-lining.

23.3. Japanese Sparrowhawk
Accipiter gularis p.98

Rather tapered wings with narrow rounded wing-tips, uniformly patterned underwing (less heavy than in Besra), and relatively short square-tipped tail with dark bars narrower than in Besra.

23.4. Chinese Sparrowhawk
Accipiter soloensis p.98

Sharply pointed wings with unmarked wing-lining, and short tail. Adult has extensive black wing-tips and, from above, unbanded tail; juv. has tail slightly banded above; most of underwing paler than body.

23.5. Nicobar Sparrowhawk
Accipiter butleri p.97

Wings look short with blunt rounded tips, white wing-lining and weak or no banding on flight feathers and tail.

23.6. Northern Goshawk
Accipiter gentilis p.99

Wings are long, broad through secondaries, with narrower hand; tail well-rounded. Adult has scarcely banded tail, and long white undertail-coverts; juv. has spotted wing-lining, bold tail-bands and whitish tail-tip (see juv. Crested).

23.7. Eurasian Sparrowhawk
Accipiter nisus p.99

In flight wings rather tapered with narrow rounded tips, strongly barred underwing, and tail relatively long and with slightly rounded tip, with usually four bands from above.

23.8. Crested Goshawk
Accipiter trivirgatus p.97

Short wings with blunt rounded wing-tips, bulging secondaries and finely speckled wing-lining; rounded tail has (usually) four dark bands, and hardly paler tip. See Jerdon's Baza (Pl. 21).

PLATE 24. *BUTEO* BUZZARDS (J. Schmitt)

Buzzards *Buteo*

Broad-winged and -tailed raptors, usually brown and heavily mottled, occurring in pale, intermediate and dark (essentially solid) morphs, making identification an extreme challenge. All have pale eyes as juveniles, darker as adults; adults also have more distinct dark trailing edge to wing and tail.

Common; montane
see text

24.1. Himalayan Buzzard
Buteo burmanicus

p. 102

Medium-sized, with tarsi more than half-feathered; dark brown belly, but often pale thighs (unlike Long-legged) and nearly unbarred dark uppertail; almost monomorphic. Adult has pale breast (most males) or variably rufous-brown breast separated from dark belly by pale U (most females). In flight, black carpal patches and dark belly-band. Adult from below has distinct broad dark trailing edge to wing and subterminal tail-band, but tail looks unbarred otherwise. See text for racial details.

Scarce; regional habitat unclear
see text

24.2. 'Steppe' Buzzard
Buteo buteo vulpinus

54cm p.101

Smallest buteo, with mostly unfeathered tarsi. From Long-legged by weaker bill, finer legs and claws and shorter-necked appearance at rest; usually more solid brown above than adult Long-legged. A buteo with heavily barred underparts will be this; rufous morph (commonest), is more rufescent than others except Long-legged, and tail is at least lightly barred. Adult has dark breast; grey-brown morph is heavily barred below, without rufous tones; rufous morph has variable rufous barring below, including wing-lining; dark morph perhaps separable from dark Long-legged only on shape. Juv. has streaked underparts (except dark morph). In flight wings relatively short, with tail slightly shorter and more rounded than in Long-legged. From below, typically has dappled or barred wing-lining often with paler secondary coverts, little or no carpal patch, and several fine dark bands (sometimes inconspicuous) on tail with stronger subterminal band; pale patch at base of primaries typically not very conspicuous. From above, tail often entirely rufous. Tail of juv. has many even-width dark bands and no strong subterminal band.

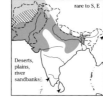

rare to S, E

Deserts, plains, river sandbanks

24.3. Long-legged Buzzard
Buteo rufinus

61cm p.102

Large with long neck, and mostly unfeathered tarsi; plumage always unbarred, and all but palest and darkest birds are rufescent, as is smaller (sometimes almost identical) 'Steppe'. Light and rufous morphs are rufous above (*vs* just edged rufous in 'Steppe') with whitish-based (often unbarred) rufous tail and rufous-brown belly-band and thighs; head and breast are pale in light morph (commonest), rufous in rufous morph. Dark morph inseparable in plumage from dark 'Steppe'. Juv. is typically non-rufescent, with dark brown back (as in 'Steppe'), and pale brown, finely barred tail with whitish base. In flight has relatively long wings, from below with dark carpal patch, usually lightly streaked wing-lining and stronger pale primary patch than 'Steppe', and rather long square-tipped tail. Rufous morph usually has paler upperwing-coverts and back than rufous 'Steppe'; dark morph often has several narrow bands and a broad dark subterminal band on tail. Juv. dark morph has broader dark bars on flight feathers and tail than dark juv. 'Steppe'.

Scarce; mountains

24.4. Upland Buzzard
Buteo hemilasius

71cm p.103

Large, with large pale lower breast-band above dark belly and long-feathered dark thighs, tarsi feathered at least three-quarters to toes, tail without strong subterminal band. Adult has often dark-hooded appearance, and greyish uppertail with somewhat wavy, narrow dark bars; light morph has lightly marked whitish head and underparts, with especially prominent dark moustachial streak. Dark morph probably not separable from dark Long-legged and 'Steppe'. Juv. typically has large pale patch on upperwing-coverts and more prominently barred tail. Some faded juvs. are mostly pale creamy, even on upperparts. In flight wings relatively long, with completely white panel on primaries; dark morph has greyer flight feathers from below than other dark buzzards; adult has unbanded tail. See text for hypothetical Rough-legged *B. lagopus*.

PLATE 25. *BUTASTUR* BUZZARDS, HONEY-BUZZARDS AND SERPENT-EAGLES (J. Schmitt)

Honey-buzzards *Pernis*

Large raptors recalling buteos except for small head with mild expression and long neck; plumage extremely variable.

Common; woods

25.1. Oriental Honey-buzzard
Pernis ptilorhyncus
65cm   p.83
Rather large; wing-tips fall short of tail-tip. 'Pigeon-headed' appearance (small bill, large eye with no overhanging brow, closely feathered face) diagnostic; short crest usually not visible. Underparts and wing-lining vary from whitish to rufous to dark brown. In flight, long-necked with long broad wings and tail. Adult has at least a hint of a dark gorget; in flight, only very tips of primaries black (less extensive than on buteos), and both wing and tail have distinctive irregular bars. Male has broad black trailing edge of wing behind broad pale band, and a broad pale central band on blackish tail; female has more bands on secondaries, and pale tail with broad subterminal band and two narrow bands on basal half. Juv. has pale head with dark eyeline, yellow cere and mottled back; in flight from above has pale primary panels, and from below dark secondaries, extensive dark primary tips, and finer tail-bands.

25.2. European Honey-buzzard
Pernis apivorus   65cm   p.83
Hypothetical; Afghanistan. Very like Oriental; see text. Adult lacks gorget or mesial stripe, almost always has dark carpal patch (not apparent in dark morph), and five 'primary fingers' rather than six. Male has broader pale band above dark trailing edge of wing; female has two irregular-width bands on secondaries (three of even width in Oriental).

Buzzards *Butastur*

Rather small, slim, narrow-winged raptors; not true buzzards.

Dry woods, fields, desert

25.3. White-eyed Buzzard
Butastur teesa
43cm   p.100
Small and slim; wing-tips fall short of tail-tip. Brown above with prominent whitish oval patch on wing-coverts, and dark mesial streak on white throat. Adult has white eye on dark head, extensive pale cere, and brown-barred underparts. Juv. has whitish head and underparts with dark eye and dark streaks. In flight, wings rather narrow with rounded tip, dark wing-lining and pale primary coverts with only tips of primaries dark, and longish tail with weak bars, broad blackish subterminal band and pale tip. From above, has pale shoulders and rufescent tail, greyer in juv. See text for similar Grey-faced *B. indicus* (p. 101), hypothetical vagrant to Andamans, Nicobars.

Serpent-eagles *Spilornis*

Small to rather large brown eagles with thick recumbent black crests, bare yellow face and legs; broad paddle-shaped wings and rounded tails are boldly banded black-and-white in most, and underparts heavily spotted with white. One widespread species and two highly localised endemics.

Endemic
Scarce; forest

25.4. Great Nicobar Serpent-eagle
Spilornis klossi
46cm   p.93
Tiny and pale, with reduced crest and unspotted ochre underparts. In flight, unmarked whitish wing-coverts, vague narrow dark bands on flight feathers, and 2–3 dark bands on tail.

Endemic
Inland forest, edge

25.5. Andaman Serpent-eagle
Spilornis elgini
50cm   p.93
Medium-sized; wing-tips fall somewhat short of tail-tip. Very dark with small white spots. In flight, wings with very broad dark bands; tail of adult has 2–3 very broad black bands, juv. has four broad black bands.

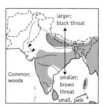
larger; black throat
Common; woods
smaller; brown throat
small, pale

25.6. Crested Serpent-eagle
Spilornis cheela
74cm   p.92
Fairly small to medium-large; wing-tips fall somewhat short of tail-tip. Juv. has white head with dark scales on crown, dark ear-patch, spangled upperparts, and whitish underparts with fine dark streaks. In flight adult has dark wing-linings, very broad dark trailing edge to wings behind broad white band, and two broad black tail-bands. Juv. has several even-width bands on under-wings and tail. Darker in Andamans, with narrower wing-bands, but paler than blackish-bellied Andaman. In C Nicobars small and pale; larger than Great Nicobar and spotted below, with two broad dark bands on tail.

PLATE 26. SHORT-TOED AND *HIERAAETUS* EAGLES (J. Schmitt)

(Snake-)eagle *Circaetus*

Large and eagle-like but with large head, dark hood and white underparts.

26.1. Short-toed Eagle
Circaetus gallicus

66cm p.92

Medium-large with bare tarsi and small feet; wing-tips reach tail-tip. Brown with owl-like head with large yellow eyes, and paler wing-coverts, and whitish below with dark breast forming bib, and dark spots or bars on belly. In flight small-bodied with long wings wide through middle and medium-length, square-cornered tail. From below, wing white with lightly marked wing-lining, and narrow bands on secondaries and inner primaries; tail is whitish with three (usually) narrow bands. Juv. usually has more rufous hood and is more lightly marked.

Eagles *Hieraaetus*

Diverse rather eagle-like raptors; group characters not obvious.

26.2. Rufous-bellied Eagle
Hieraaetus kienerii

57cm p.108

Small with short crest, black upperparts, and white throat and breast; wing-tips almost reach tail-tip. Adult has dark rufous belly, vent and wing-lining; juv. is entirely white below and on face and brow. In flight, wings are relatively long and narrow; flight feathers and tail pale from below and lightly barred. From above, all dark with browner primaries; viewed head-on, leading edge of wing is white.

26.3. Booted Eagle
Hieraaetus pennatus

52cm p.108

Small and compact, in three colour morphs; wing-tips fall short of tail-tip when perched and tarsi fully feathered. Dark brown above with paler crown, nape and wing-coverts, and white, dark brown or rufous below and on wing-lining. In flight, shows (head-on) two white shoulder-spots at leading edge of wing; from above, paler bar on wing-coverts and whitish U on uppertail-coverts; from below, densely barred flight feathers with paler wedge on inner primaries, pale tail usually with dusky subterminal band and, sometimes, faint bands. Wing and tail have uneven-width white trailing edge.

26.4. Bonelli's Eagle
Hieraaetus fasciatus

70cm p.107

Rather large and pale with heavily feathered legs and wing-tips that fall short of tail-tip. In flight has large head and long neck, long, rather pointed broad wings with secondaries darker than primaries, and long squarish pale and narrowly barred tail. Adult is dark brown above with whitish mantle (forming triangle), and white with black streaks below; in flight has black band on wing-lining, finely barred primaries, and wide subterminal band on tail. Young adults have entirely barred flight feathers with broad dark trailing edge, obscured by dark in older birds. Juv. is paler above and rufescent below with fine streaks, fading sometimes to whitish by spring; in flight has lightly marked wing-linings, sometimes with dark band formed by tips of greater coverts, and finely barred flight feathers and tail. Second-winter (imm.) more heavily streaked below.

PLATE 27. HAWK-EAGLES AND BLACK EAGLE (J. Schmitt)

Hawk-eagles *Spizaetus*

Sturdy with somewhat paddle-shaped wings with trailing edges pinched in at body; adults have dark mesial streaks and pale eyes. Note facial and underparts patterns (especially belly), wing and tail patterns in flight.

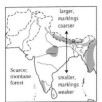

27.1. **Mountain Hawk-eagle**
Spizaetus nipalensis

72cm p.109

Much as Changeable and Crested but larger, with large bill and claws; wing-tips reach halfway down tail and tarsal feathering runs onto toes. Note the long white-tipped crest and largely pale wing-coverts. In flight, wings short, square-tipped, pinched-in at body, and heavily banded below and on tail. Adult has dark cheek, heavily streaked breast and banded flanks and belly. Juv. pale, with spots (vs streaks) on crown and nape. Smaller in *kelaarti* of S Peninsula and Sri Lanka, with rufous bars below.

27.2. **Changeable Hawk-eagle**
Spizaetus limnaeetus

71cm p.108

Mid-sized with short crest; wing-tips reach two-thirds down tail, and tarsal feathering extends only to base of toes. In flight wings more round-tipped than Mountain, and has dark morph (not true of Crested). Adult has lightly streaked head without dark cheek-patch and lacks heavy barring on belly of Mountain; darker above without markedly pale wing-coverts; weakly and narrowly barred secondaries. In flight, adult wings and tail are narrowly barred, both with wider subterminal band; uneven spacing of tail-bands recalls Oriental Honey-buzzard (Pl. 25). Dark morph (rare, SE) adult is all dark with unbanded paler undertail and wing, with darker wing-tips and subterminal tail-band. Juv. has buffy-white virtually unmarked head and underparts and may have dark eyes; in flight has darker primaries and evenly spaced narrow tail-bands. Juv. dark morph said to have barred wings and tail.

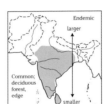

27.3. **Crested Hawk-eagle**
Spizaetus cirrhatus

71cm p.109

Structurally similar to Changeable, but has very long, white-tipped black crest at all ages, and lacks dark morph. Looks more variegated and streaked above than Changeable. Adult has streaked throat, broadly streaked breast, dark belly-patch (heaviest on female and prominent in flight) and unbanded vent.

Eagle *Ictinaetus*

Large but slim, broad-winged, all-dark kite-like raptor. Skims above forest canopy.

27.4. **Black Eagle**
Ictinaetus malayensis

75cm p.104

Large, black and kite-like with bright yellow cere and legs; wing-tips reach or exceed tail-tip. In flight, large paddle-shaped, long-fingered wings and long, faintly banded tail. Juv. blackish-brown, often with vague paler marks on neck and underparts.

Plate sponsored by Mr. David Rockefeller.

PLATE 28. *AQUILA* EAGLES PERCHED (J. Schmitt)

See Plate 29 for flight illustrations

Eagles *Aquila*

Generally large brown eagles, best identified both on structure and plumage, especially in the rather plain adults. Separable into two groups: the smaller spotted eagles have rounded nares and short-feathered tarsi; the others have oval nares on heavier bills, and full-feathered tarsi with larger feet.

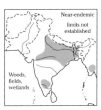
Near-endemic
limits not established
Woods, fields, wetlands

28.1. Indian Spotted Eagle
Aquila hastata

64cm   p.105

Mid-sized, with smallish bill, long gape and small feet. Adult is drab brown and unspotted, with yellowish-brown eyes and often paler brown vent and tarsi. Juv. and first-summer have brown eyes and pale flecks on nape (no rufous patch), upper back and wing-coverts, larger pale spots on median coverts and narrow pale tips to tertials; and are often heavily streaked below.

Common; dry woods, fields

28.4. Tawny Eagle
Aquila rapax

67cm   p.105

Medium-large; wing-tips reach tail-tip. More lightly built than Steppe, with gape-line to mid-eye, thinly feathered pale lores below reduced brow-ridge, and mid-sized, more extensively pale bill. Variable: whitish to dull rufous- or greyish-brown to (rarely) dark brown with paler vent and leg feathers and pale streaks on belly; or entirely streaked below. Often dark-crowned or -hooded; and/or blotched and streaked overall, but never has rufous nape of Steppe. Adult has yellow eyes. Juv. has brown eyes, and is heavily spangled grey-buff above, often a paler-tipped hindcollar, narrow pale tips to greater coverts and secondaries.

Wetlands

28.2. Greater Spotted Eagle
Aquila clanga

67cm   p.105

Mid-sized with rather small bill and long spiky nape feathers; usually darker than Indian, with brown eyes, longer thigh feathers and white lower tarsal feathers. Juv. usually black, blackish-brown or dark rufous-brown, with dark eyes, large pale spots on wing-coverts, and pale streaks below; older immatures blotchy. Rare pale morph immature ('*fulvescens*') is tawny to rufous-tawny, typically with pronounced pale spotting above (unlike Tawny).

Montane forests, open slopes

28.5. Golden Eagle
Aquila chrysaetos

95cm   p.107

Very large with huge feet and claws; wing-tips fall short of tail-tip. Dark brown with amber to yellow eye, golden crown and nape and pale-mottled wing. Juv. dark brown with dark iris.

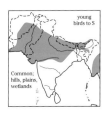
young birds to S
Common; hills, plains, wetlands

28.3. Steppe Eagle
Aquila nipalensis

78cm   p.106

Large and heavy; wing-tips reach tail-tip. Dark and blotchy with rufous nape-patch (and often pale chin), yellow gape reaching rear edge of (brown) eye, dark feathered lores below heavy brow, large bill, very heavily feathered tarsi, large feet and claws. Has rare rufous-brown morph. Juv. often paler greyish-brown above with broad white tips to greater wing-coverts and secondaries (as in Greater Spotted), and unstreaked below; immature in third summer can resemble adult Tawny and Indian Spotted.

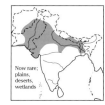
Now rare; deserts, wetlands

28.6. Eastern Imperial Eagle
Aquila heliaca

85cm   p.106

Large with medium-sized feet; wing-tips fall near tail-tip. Blackish with very pale nape-patch more strongly contrasting than in Golden, blackish chin and white scapular spots. Juv. dark above with pale crown and pale streaks on mantle and wing-coverts, and buff-white below with dark streaks on breast and belly. Second-year mottled, with dark throat and pale crown.

PLATE 29. *AQUILA* EAGLES IN FLIGHT (J. Schmitt)

See Plate 28 for non-flight illustrations

Eagles *Aquila*

Spotted eagles have white crescent at base of outer primaries and no dark trailing edge to wing, unlike the larger species.

29.1. Indian Spotted Eagle
Aquila hastata p.105

In flight wing slightly narrower than Greater, with slightly pale wing-lining and finely and vaguely barred flight feathers and tail, but no dark trailing edge or subterminal band on either. Relatively longer-tailed than Greater. See Steppe (which has larger bill, more rounded trailing edge of wing, more heavily banded flight feathers, with dark trailing edge in adults).

29.2. Greater Spotted Eagle
Aquila clanga p.105

In flight patterned like Indian, but wing-lining darker than flight feathers, and very broad wings make tail look short. Head, wings and tail appear relatively shorter than larger *Aquila* species. Pale morph has broader wings and blacker flight feathers and tail than Tawny.

29.3. Steppe Eagle
Aquila nipalensis p.106

Looks short-necked with long, coarsely barred wings, long primary 'fingers' and medium-length, rather wedge-shaped barred tail with dark subterminal band; from above has pale base to inner primaries and often a white central back-spot. Adult has rectangular wings with greyish flight feathers, black carpal patch, and dark trailing edge to wing. Juv. has narrower wings with S-curved white-tipped trailing edge, broad white band through underwings, and white uppertail-coverts, vent and tail-tip.

29.4. Tawny Eagle
Aquila rapax p.105

Looks longish-necked with shorter, narrower 'hand' than Steppe, vaguely barred secondaries, no dark carpal patch and more rounded tail. Pale morph often has pale triangle on secondaries (*vs* entirely dark flight feathers of '*fulvescens*' Greater Spotted). Dark morph has more vaguely barred underwing than Steppe, without a darker trailing edge.

29.5. Golden Eagle
Aquila chrysaetos p.107

Broad wings with strongly curved trailing edge, and slightly rounded longish tail. Adult from above has tawny bar on greater secondary coverts; from below, looks dark with greyish marbling on flight feathers and tail, both with broad darker trailing edges. Juv. has large white panel on base of primaries, and white base of tail.

29.6. Eastern Imperial Eagle
Aquila heliaca p.106

Longer head and neck extension than Golden, with straighter trailing edge to wings, and medium-length, squarer-cornered tail. Adult lacks pale bar on upperwing-coverts, has broader dark trailing edge on wing and tail, and blackish wing-lining. Juv. pale, especially on head, with striped wing-lining and body, pale triangle on inner primaries, and unmarked pale thighs and vent.

Plate sponsored by Mr. Perry Bass.

PLATE 30. OSPREY, FISH-EAGLES AND SEA-EAGLES (J. Schmitt)

Osprey Pandionidae

Large, eagle-like fish-eating raptor.

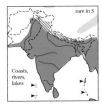

30.1. Osprey
Pandion haliaetus

56cm   p.111
Dark above and white below with blackish eye-stripe on white head; in flight, long wings held crooked, with white linings and dark carpal and wing-tips. Juv. scaled white above.

Fish-eagles *Ichthyophaga*

Grey eagles with white bellies, rather pale heads and broad dark wings; thighs short-feathered, unlike sea-eagles.

30.2. Lesser Fish-eagle
Ichthyophaga humilis

64cm   p.87
Rather small; wing-tips nearly reach tail-tip. In flight, from larger Grey-headed by dusky base of shorter tail. Juv. is pale with scarcely streaked head and virtually unstreaked underparts (unlike juv. Grey-headed); in flight, marbled rather than banded flight feathers.

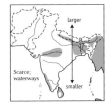

30.3. Grey-headed Fish-eagle
Ichthyophaga ichthyaetus

74cm   p.88
Mid-sized; larger and longer-tailed than Lesser, with wing-tips falling well short of tail-tip. In flight, white tail with broad black tip. Juv. is streaked overall except on belly and vent, with no dark cheek-patch (unlike juv. sea-eagles); in flight, white underwings with lightly barred flight feathers and tail.

Sea-eagles *Haliaeetus*

Large and generally pale-headed with full-feathered thighs, unlike *Ichthyophaga* fish-eagles. Juveniles are browner and streaked, with dark eyelines or cheek-patches.

30.4. White-bellied Sea-eagle
Haliaeetus leucogaster

68cm   p.86
Long-necked and small-headed, with long narrow wings and wedge-shaped tail. Adult is white with grey mantle and wings; in flight, white with dark flight feathers and base of tail. Juv. is dark brown with pale head and belly; in flight, note dark breast-band and flight feathers, with large pale panel on base of primaries and white tail with broad dark tip.

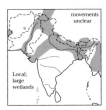

30.5. Pallas's Fish-eagle
Haliaeetus leucoryphus

80cm   p.86
Dark with a pale head, dark wings and very broad white band on tail. Juv. mottled above with dark cheek on paler head, and paler underparts; in flight, has whitish wing-lining and inner primaries with darker leading edge, and solid dark secondaries and tail.

30.6. White-tailed Eagle
Haliaeetus albicilla

77cm   p.87
Brown with mottled foreparts and whitish head; in flight, long, broad dark wings and short, largely white wedge-shaped tail. Juv. and imm. are dark and heavily mottled, with pale axillaries and whitish tail with dark-edged feathers.

PLATE 31. *GYPS* VULTURES (J. Schmitt)

Vultures *Gyps*

Mid-sized to very large brownish vultures lacking the ornamental features (plumage and facial) of other vultures, apart from a pale neck-ruff. Often difficult to identify, particularly juveniles. Note bill colour and structure, colour and feathering of head and neck; in flight, underpart patterns.

Now local; open areas, villages, cities

31.1. White-rumped Vulture
Gyps bengalensis

85cm   p.89

Size of Indian and Slender-billed but black with white neck-ruff, grey secondaries, and clear white rump; shorter, stouter bill is dark with silvery upper mandible. In flight, black with striking white wing-lining, dark leading edge to wing, and dark vent (not contrasting with tail). Juv. is entirely dark; darker than juv. Indian and Slender-billed, less streaked above but more sharply below, with dense white down on face and neck (brownish on nape) and blackish bill. In flight all dark below with pale neck, whitish bars in wing-lining, darker leading edge and axillaries.

W, S limits unknown

31.3. Slender-billed Vulture
Gyps tenuirostris

92cm   p.90

Mid-sized and similar to Indian but generally dingier above, with small, ragged neck-ruff, narrower pale edges to tertials and greater wing-coverts, pale but mottled rump and less full-feathered thighs, belly and vent. Head and thin neck are always black with elongate bill, dark cere, low forehead and peaked crown; at close range, note oval nostrils and virtual absence of head feathering. Bill typically has only culmen-ridge pale (but may be more extensive), and claws are dark. Juv. has only slightly more down on head and neck. In flight note moderate contrast between greyish body and dark flight feathers and no contrast between neck and breast; white thighs and dark undertail-coverts. Wings are less even-width than Indian's, with bulging secondaries; tail narrow.

Endemic
N and E limits unknown

Now rare; dry woods, cliffs, plains

31.2. Indian Vulture
Gyps indicus

92cm   p.90

A mid-sized tawny vulture with full thick white neck-ruff, broad pale edges to dark tertials and greater coverts, and white rump. From very similar Slender-billed by more rounded head and thicker neck (both not as black), with stouter pale yellowish bill with pale cere, and pale claws. At close range, note slit-like nostril and fine white hair-like feathers on head. Thighs, belly and vent more fully feathered, giving hindquarters heavier look. In flight uniformly buff below, with dark flight feathers and strong neck/breast contrast; wings are broad and straight and tail broad. Juv. has a dark bill with pale culmen, similar to adult Slender-billed, but has much down on head and neck; neck is initially pinkish.

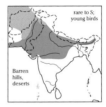

Barren hills, deserts

rare to S; young birds

31.4. Griffon Vulture
Gyps fulvus

100cm   p.91

Large and often rufescent, with down-covered head, pale eye and bill, dark cere and legs, and white ruff. Most like Indian, but larger and often more rufous with a paler, downier head, less sharply pale rump. In flight, whitish wing-lining with tawny axillaries and leading edge, dark rear trim and flight feathers. Juv. is similar but vaguely streaked above and more prominently streaked below, with dark bill and eye.

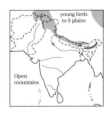

young birds to S plains

Open mountains

31.5. Himalayan Vulture
Gyps himalayensis

120cm   p.90

Huge and very pale, full-feathered with downy head and pale bill; whitish ruff does not contrast with body. In flight, looks white with black flight feathers and tail. Juv. is very dark with crisp pale streaks overall and partly pale bill; in flight, dark with pale head and fine white stripes.

PLATE 32. OTHER VULTURES (J. Schmitt)

Vultures *Neophron, Gypaetus*

Slim, wedge-tailed vultures, pale-bodied as adults, dark as juveniles. The two species are otherwise very different.

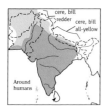

32.1. Egyptian Vulture
Neophron percnopterus

64m   p.89

Small with frilly nape, bare face and thin bill, and wedge-shaped tail. Adult is white with black flight feathers. Juv. is dark with pale vent and tail.

32.2. Bearded Vulture
Gypaetus barbatus

122cm   p.88

Huge, with long wings and long wedge-shaped tail. Adult has buffy head and underparts. Juv. is dark with pale streaks except on head; appears paler-bodied and longer-tailed than Cinereous, with less rounded wing-tips.

Vultures *Aegypius*

Large vultures with heavy, blackish-plumaged bodies and ornamented heads.

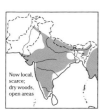

32.3. Red-headed Vulture
Aegypius calvus

84cm   p.91

Mid-sized and dark with heavy bill, naked reddish head, inner thigh-spots and feet, and shortish rounded tail. Male has pale iris and grey tertials; female has dark iris and whitish tertials. Adult is black with paler secondaries; in flight, dark with striking white flank-patches and pale band through base of flight feathers. Juv. is browner with pale belly and vent; in flight, has more bulging secondaries.

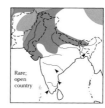

32.4. Cinereous Vulture
Aegypius monachus

105cm   p.91

Huge and dark, with thick ruff, downy crown, pale occiput and base of heavy bill, and pale feet. Adult has pale crown, ruff and upper breast, dark in juv. In flight, wings very broad and rectangular; wedge-shaped tail is shorter than in Bearded.

PLATE 33. HARRIERS I (J. Schmitt)

Harriers *Circus* (continued on Plate 34)

Rather small, long-winged and long-tailed, lightly built raptors of open areas. All species are sexually dimorphic, and often extremely difficult or impossible to identify with certainty owing to pronounced and complex age variation. Note especially wing and tail patterns.

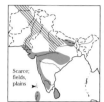

Scarce; fields, plains

33.1. Montagu's Harrier
Circus pygargus

48cm　　　　　　　　　　p.96

Rather delicate with a very small bill; shorter-legged than Pied, with more horizontal stance. Wing-tips reach tail-tip; in flight has narrow wing-tip and narrow white uppertail-covert band. Male is grey with heavy rufous streaks on lower breast and belly; in flight, grey wings with darker secondary coverts, black bar on secondaries and dark wing-tip, and grey tail with blackish and rufous bands. From below, rufous wing-lining. Female is brown above with whitish face and indistinct rim to facial disk, and heavily streaked buffy underparts. In flight, female underwing has complete white band on dark secondaries, and dark tips on inner primaries. Juv. is dark brown with rufous crown, a faint narrow rufescent face-ring, buff patch on wing-coverts, and rufous-buff underparts, with few streaks on breast and sides. In flight, juv. has rufous underwing-coverts, somewhat barred axillaries, more extensive dark wing-tip and dark trailing edge than juv. Pallid, but all characters variable.

Common; dry fields, wetlands

33.2. Pallid Harrier
Circus macrourus

48cm　　　　　　　　　　p.95

Very similar to Montagu's (narrow wing-tip and narrow white uppertail-covert band), but with longer legs and more upright stance; wing-tips fall short of tail-tip. Male is very pale grey with black wing-tips (with pale inner primaries visible at rest) and white belly. In flight, very pale with black wedges on wing-tip, and dark grey bands on tail. Female from female Montagu's by more conspicuous narrow pale rim to facial disk (complete around throat), and usually darker cheek-patch; thighs diffusely spotted rufous (*vs* finely streaked). In flight, from below has dark secondaries and paler primaries, with a pale band on secondaries that does not reach body, and a pale trailing edge to inner primaries. Juv. from juv. Montagu's by more obvious facial rim, bolder dark face-patches and typically unstreaked yellowish-buff underparts. In flight, often indistinguishable from juv. Montagu's; little-marked axillaries. See text for subadult plumages.

rare to W, S

Grass, fields, wetlands

33.3. Pied Harrier
Circus melanoleucos

48cm　　　　　　　　　　p.95

Mid-sized with relatively long tail and legs; wing-tips nearly reach tail-tip. Note complete facial disk; in flight, rounded wing-tip and and white rump-band. Male sharply pied: pale with black hood and mantle; in flight, note white forewing enclosed by black border and large black wing-tip. Female is greyer than any other female or juv. harrier, with fine black streaks on whitish collar, paler shoulder, finely streaked breast and whiter belly. In flight from above, female has pale grey, crisply banded flight feathers and tail, both with broad subterminal band; from below, wings whitish with subterminal band and narrowly banded secondaries. Juv. is solid dark brown above, and heavily streaked rufous-brown below with narrow dark-and-light-streaked throat-collar. In flight, dark rufous wing-lining and banded primary bases; greyish tail with broad dark bands but little contrast above. See text for subadult plumages.

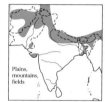

Plains, mountains, fields

33.4. Hen Harrier
Circus cyaneus

50cm　　　　　　　　　　p.94

Mid-sized, with rounded wing-tip not nearly reaching tail-tip. In flight has broad rounded wing-tips and broad white uppertail-covert band. Male is pearly-grey with grey head and upper breast (giving hooded look unlike male Pallid) and white below; all primaries visible in folded wing black. In flight, sharply cut-off black wing-tips, and grey tail with faint dusky bands; white underwings with black trailing edge. Female is brown above with broadly streaked hindneck and not very prominent pale face pattern with somewhat streaked ear-coverts, and heavy brown streaks below turning into teardrops on belly, thighs and vent; in flight, all flight feathers boldly barred. Juv. like female but with bolder face pattern, entirely streaked and often buffier underparts; usually duller below with less well-marked face pattern and more broadly streaked hindneck than juv. Pallid or Montagu's. In flight, very like female, but undersides of secondaries darker and more uniform. See text for subadult plumages.

PLATE 34. HARRIERS II (J. Schmitt)

Harriers *Circus* (continued from Plate 33)

Common; wetlands

34.1. Western Marsh Harrier
Circus aeruginosus

56cm   p.94

Large and sturdy, with wings not reaching tail-tip; usually lacks white band on uppertail-coverts and most plumages have whitish head with wide dark eye-stripe. Male is dark rufous-brown with pale, lightly streaked head, and pale grey unbanded wing and tail; belly may be whitish. See text for dark-morph male. In flight, male from above is tricoloured, with solid dark brown back and secondary coverts, pale grey secondaries and primary coverts, and black tips to outer primaries; underwing whitish with black wing-tip. Female blackish-brown with buffy-white crown, throat, breast-band and patch on wing-coverts, and rufous tinge to belly, vent and tail. In flight, dark with a little pale in primary bases and pale leading edge of wing. Juv. similar to female but without rufous; may be entirely dark with only nape pale. In flight, very dark, underwing with unbanded primaries and only small whitish carpal crescent.

34.2. Eastern Marsh Harrier
Circus spilonotus   56cm   p.94

Hypothetical; mainly N. Large and sturdy, with wings not reaching tail-tip; usually has white band on uppertail-coverts and lacks dark eye-stripe of juv. and female Western. Male recalls male Pied (Pl. 33; adults here for comparison), but lacks crisp contrast, with black crown and mantle, black streaks on hindcollar and breast, and white scales on blackish upperparts. Male in flight has wings speckled black above with dark lesser coverts (*vs* pale in male Pied). Female is rufous-brown, streaked on head and underparts, and mottled above. Female in flight has flight feathers and tail dark grey above with dark bands; below, secondaries dark grey next to whitish bases of primaries and often a dark carpal patch. Juv. is blackish-brown with pale face, breast-band and patches on wing-coverts, and rufous-brown belly; in flight, bases of primaries whitish and unbanded, and uppertail-coverts can be dark brown or whitish. See text for subadult.

PLATE 35. FALCONETS, MERLIN AND KESTRELS (J. Schmitt)

Falconets *Microhierax*

Tiny, big-headed shrike-like falcons with black cheek-patches and upperparts, and white underparts.

Forest edge

35.1. Collared Falconet
Microhierax caerulescens

18cm p.111

Rufous throat, belly and vent. In flight, white wing-lining and barred flight feathers with black trailing edge; tail black above, barred below.

Forest edge, gardens

35.2. Pied Falconet
Microhierax melanoleucos

20cm p.111

Slightly larger than Collared, with black flanks; no rufous. In flight, similar to Collared.

Falcons *Falco* (continued on Plates 36–37)

Compact birds of prey with generally long pointed wings and longish narrow tails; most with narrow dark vertical moustache. Flight is strong and direct, although some hover while hunting. Several species are highly variable in plumage. Note facial pattern, wing and tail pattern and length of wings when perched; also pattern of markings on underparts.

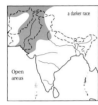

a darker race

Open areas

35.3. Merlin
Falco columbarius insignis

29cm p.114

35.4. *Falco columbarius pallidus*

Compact; wing-tips fall well short of shortish tail (just reaching black subterminal band). Sexes very different, but both have diffuse facial pattern with pale streaked hindcollar and narrow brow, streaked underparts, and broad dark subterminal tail-band. Male is silvery-grey above with darker primaries, and finely streaked on buff underparts and hindcollar. Female and juv. are brown above with broad dark bands on tail, and heavily streaked below. In flight, short-winged, with barred underwing. Race *insignis* is much darker than scarcer *pallidus*.

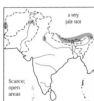

a very pale race

Scarce; open areas

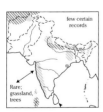

few certain records

Rare; grassland, trees

35.5. Lesser Kestrel
Falco naumanni

34cm p.112

Lightly built, with long wing-tips reaching into broad black subterminal tail-band. Moustache weak or absent, and claws pale (dark in Common). Male has plain grey head and unspotted chestnut mantle, grey greater wing-coverts and tail, and buffy underparts with light spots. Female and juv. from female Common by lack of dark line behind eye, and paler, less streaked cheek. In flight, looks short-bodied with wedge-tipped tail; wing-tip more pointed than in Common. Male has strikingly pale underwing and tail with lightly spotted wing-lining; above, wing rufous-and-grey with black tip. Female in flight from female Common by paler, less marked bases of underside of flight feathers.

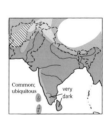

Common; ubiquitous

very dark

35.6. Common Kestrel
Falco tinnunculus

36cm p.112

Medium-sized; wing-tips fall short of black tail-band. Always has long narrow moustache. Male has black-spotted rufous upperparts, no grey on wings, and pale buff underparts with streaks on breast and spots on belly. Female and juv. are rufous-brown above with close barring; from female Lesser by dark line behind eye. In flight, wing-tips rather rounded; underwing evenly dark-marked.

PLATE 36. MID-SIZED FALCONS (J. Schmitt)

Falcons *Falco* (continued from Plate 35 and on Plate 37)

Compact birds of prey with generally long pointed wings and longish narrow tails; most with narrow dark vertical moustache. Flight is strong and direct, although some hover while hunting. Several species are highly variable in plumage. Note facial pattern, wing and tail pattern and length of wings when perched; also pattern of markings on underparts.

Local; forest edge, fields, cliffs

36.1. Oriental Hobby
Falco severus

29cm   p.115

Rather small; wing-tips exceed tail-tip. Black above and rufous below with white throat and half-collar. See much larger *peregrinator* race of Peregrine (Pl. 37), which often has barred flanks. Juv. is streaked and spotted with black below. In flight looks short-tailed; note rufous wing-lining, plain in adult, streaked and/or barred in juv.

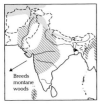

Breeds montane woods

36.2. Eurasian Hobby
Falco subbuteo

33cm   p.114

Rather small; wing-tips just exceed tail-tip. Dark grey above with two short moustaches; white below with streaked breast and flanks, and rufous thighs and vent. Juv. is browner above, and streaked below on thighs and vent. In flight, narrow-winged and shortish-tailed; underwings look all dark and undertail finely barred.

Presence uncertain

Rocky coasts

36.3. Sooty Falcon
Falco concolor

38cm   p.114

Hypothetical; SW Pakistan. Mid-sized; wing-tips extend far beyond tail-tip. Solid sooty-grey with yellow soft parts. Juv. is brown above and buffy below with dark spots; face pattern like Eurasian Hobby's. In flight, wings very long and narrow; juv. has darker underwings than Eurasian Hobby, tail narrowly pale-barred below with broad pale tip.

36.4. Eleonora's Falcon
Falco eleonorae   40cm   p.115

Hypothetical. Slender and long-winged; wing-tips just exceed tail-tip. Dark morph is entirely blackish; pale morph recalls Eurasian Hobby, but with single dark moustache and more extensively rufous underparts. See text for juv. In flight, very dark wing-lining with paler flight feathers, and long narrowly barred tail.

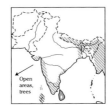

Open areas, trees

36.5. Amur Falcon
Falco amurensis

30cm   p.113

Small, compact; wing-tips reach or exceed tail-tip. Male is dark grey with rufous thighs and vent; in flight, striking white wing-lining and black flight feathers. Female recalls Eurasian Hobby, but is barred above and below with buffy thighs and vent; in flight, heavily barred underwings with broad dark trailing edge, and pale grey tail with narrow black bars and dark subterminal band. Juv. like female but scaled with buff above, streaked black below; in flight, narrower dark trailing edge, and no broader subterminal tail-band.

Now scarce; plains, groves, villages

36.6. Red-headed Falcon
Falco chicquera

34cm   p.113

Wing-tips fall far short of tail-tip. Pale grey with fine dark barring, including tail with broad black subterminal band; note rufous crown, nape and moustache. Juv. has some black streaks on crown and rufous on upper back.

36.7. Red-footed Falcon
Falco vespertinus   35cm   p.113

Hypothetical; Afghanistan. Male identical to male Amur, but with dark grey wing-lining. Female is grey above with black bars, and unmarked orange-buff head and underparts; in flight, and unmarked buff wing-linings, and black-barred flight feathers. Juv. as juv. Amur but often with browner streaks below.

PLATE 37. LARGE FALCONS (J. Schmitt)

Falcons *Falco* (continued from Plates 35–36)

Compact birds of prey with generally long pointed wings and longish narrow tails; most with narrow dark vertical moustache. Flight is strong and direct, although some hover while hunting. Several species are highly variable in plumage. Note facial pattern, wing and tail pattern and length of wings when perched; also pattern of markings on underparts.

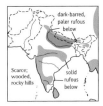

37.1. Peregrine Falcon (Shaheen)
Falco peregrinus peregrinator

43cm   p.116

Very dark grey above (almost black) with blackish hood and broad dark moustache, and rich rufous underparts; when perched, wing-tips reach tail-tip. In N, underparts are paler and more barred. See smaller but similarly coloured Oriental Hobby (Pl. 36).

37.2. Peregrine Falcon
Falco peregrinus calidus

43cm   p.116

Grey above with blackish hood and broad dark moustache, and barred with black on white underparts; when perched, wing-tips reach tail-tip. Juv. is brown above with pale brow and nape, and heavily streaked below, spot-like on flanks. In flight, underwings are densely barred and look uniformly dark, but without darker axillaries. Juv. has heavily barred undertail-coverts.

37.3. 'Red-capped' Falcon
Falco peregrinus babylonicus

40cm   p.116

Smaller and paler than *calidus* race of Peregrine, with pale rufous crown and nape, pale brow, and narrower moustache; pale salmon wash below with light barring on flanks. Juv. is duller with weakly patterned brown head and buffy underparts with fine streaks. In flight, heavy, short-winged; adult has pale, finely spotted underwings, with salmon coverts, tail-bars darkest and broadest near tip; juv. has pale buff underwings.

37.4. Laggar Falcon
Falco jugger

45cm   p.115

Heavily built with wing-tips almost reaching tail-tip. Brownish-grey above with dull rufous crown, narrow white brow, bold black eyeline and nape, long narrow moustache and white cheeks; breast paler than belly and flanks; white below with dark belly-band. Juv. is very dark, with same head pattern but heavy streaks on cheek and nearly black underparts. In flight, wing-lining dark, contrasting with paler flight feathers. Adult has grey tail with narrow pale bands and pale tip; juv. has unbanded dark tail.

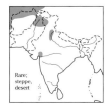

37.5. Saker Falcon
Falco c. cherrug

52cm   p.115

Large and heavy, with wing-tips falling well short of tail-tip. Brown with pale brow, and weak moustache; white below with spotted breast and heavily streaked belly and flanks. Nominate race has paler crown, and weaker facial pattern. Juv. is heavily streaked below. In flight, wings long and broad-based: two-toned above; pale below with darker central wing-lining. Tail long, pale and narrowly barred.

37.6. 'Eastern' Saker Falcon
Falco cherrug milvipes

52cm   p.115

From other races of Saker by darker crown, moustache and cheeks, rufescent barring above, and barred vs streaked flanks and wing-lining.

PLATE 38. PARTRIDGES AND FRANCOLINS (J. Schmitt)

Partridges *Alectoris, Ammoperdix, Perdix*

Mid-sized, rather pale galliformes with striped flanks. Deserts and/or mountains. Darker species on Plate 39.

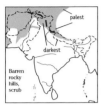
Barren rocky hills, scrub

38.1. Chukar Partridge
Alectoris chukar

38cm   p. 120

Medium-sized with red bill, legs and eye-ring; pale with white throat-patch enclosed by striking black band from forehead to breast, and black-and-rufous bars on flanks. In flight, tail rufous. Juv. much duller, more mottled, lacks distinctive marks.

Dry hills, dunes

38.2. See-see Partridge
Ammoperdix griseogularis

26cm   p. 119

Smallish and pale with orange bill and yellow legs. Male has black brow and whisker-mark, white eye-stripe, pale grey throat and chestnut-and-black stripes (not bars) on flanks. Female is

Alpine scrub, steppe, fields

plain sandy-grey with whitish brow and lightly mottled head and breast. In flight, similar to larger Chukar.

38.3. Tibetan Partridge
Perdix hodgsoniae

31cm   p. 121

White face, throat and brow, black cheek-patch, broad chestnut collar and bars on flanks, and close black barring on breast and belly; juv. lacks rufous collar. In flight, note rufous tail and rufescent primaries. See text for hypothetical Daurian *P. dauurica* (p. 123).

Francolins *Francolinus*

Mid-sized, densely patterned partridges of generally open scrubby areas; most species are widespread and noisy. Only Black is markedly dimorphic. Note facial pattern, pattern of underparts.

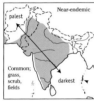
Common; grass, scrub, fields

38.4. Grey Francolin
Francolinus pondicerianus

33cm   p. 121

Pale-faced and almost entirely barred. Buff face is paler than Painted's, and lacks rufous vent. In flight, tail conspicuously chestnut and primaries dark.

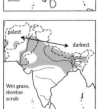
Wet grass, riverine scrub

38.5. Black Francolin
Francolinus francolinus

34cm   p. 120

Male has black face and underparts with white cheek-patch, chestnut collar and heavy white spots on flanks. Female from Painted by chestnut nuchal collar and dark cheek-stripe on rufous-buff face. Juv. similar to female. In flight similar to Painted.

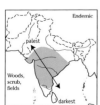
Woods, scrub, fields

38.6. Painted Francolin
Francolinus pictus

31cm   p. 120

From female Black by brighter rufous face and throat without dark cheek-stripe, and more heavily white-spotted underparts. In flight, very like female Black, but primaries more rufescent.

Scrub, field edges

38.7. Chinese Francolin
Francolinus pintadeanus

33cm   p. 121

White face with dark moustachial stripe and rufous lateral crown-stripe. Male is black, boldly spotted and barred with white with chestnut scapulars and vent. Female is browner, and mostly barred below. In flight, blackish flight feathers with white bars, and black tail nearly covered by barred coverts.

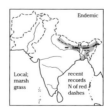
Local; marsh grass

38.8. Swamp Francolin
Francolinus gularis

37cm   p. 121

Large and long-legged with whitish lores and brow, rufous throat, and white streaks on brown underparts. In flight, bright rufous primaries and tail.

PLATE 39. PARTRIDGES, HILL-PARTRIDGES, HIMALAYAN QUAIL AND MEGAPODE (J. Schmitt)

Partridges *Bambusicola, Lerwa*

Mid-sized, rather dark galliformes with strongly marked flanks. Hills or alpine zone.

Common; hill grass, scrub

39.1. Mountain Bamboo Partridge
Bambusicola fytchii

35cm   p. 128

Long-tailed; rufous head with pale brow, face and throat, chestnut streaks on foreparts, and large black spots on flanks. Male has black postocular stripe. In flight, flight feathers and tail chestnut.

Alpine pastures, scrub

39.2. Snow Partridge
Lerwa lerwa

38cm   p. 118

Large with bright red bill and red legs. Barred black-and-white above (appearing grey) with chestnut-and-white streaks below. In flight, blackish primaries and white trailing bar on secondaries, narrower than on larger Tibetan Snowcock; tail entirely barred.

Hill-partridges *Arborophila*

Rotund forest partridges, boldly patterned in grey, brown and chestnut. Sexes similar or moderately different, and juveniles only slightly different. Note facial pattern, colour of breast.

Common; montane evergreen forest; more white on neck, more chestnut above; paler, darker

39.3. Common Hill-partridge
Arborophila torqueola

28cm   p. 126

From other hill-partridges by blackish legs. Male has black head with rufous crown and cheek, and white gorget. Female from Rufous-throated by pale lores, buffier face, and black scales on olive-grey upperparts. In flight, dark primaries and rufous-chestnut secondaries.

Foothills evergreen forest, bamboo

39.5. White-cheeked Hill-partridge
Arborophila atrogularis

28cm   p. 127

Black mask and throat with white brow and face patch; lacks rufous. Upperparts barred, like Common.

Local, scarce; evergreen forest, bamboo; Endemic

39.4. Chestnut-breasted Hill-partridge
Arborophila mandellii

28cm   p. 128

Rather dark with chestnut head and breast, broad pale postocular brow and white-and-black gorget; variable white submoustachial streak. Belly darker than other hill-partridges.

Evergreen forest, regrowth; throat blacker

39.6. Rufous-throated Hill-partridge
Arborophila rufogularis

27cm   p. 126

From Common by broad whitish brow and submoustachial streak, black-spotted rufous hindcollar, unscaled upperparts and red legs; from female Common by darker lores, greyer breast. Black-throated with rufous on upper breast in S Assam hills.

Quail *Ophrysia*

Distinctive, mysterious and quail-sized but with a longer tail. Known only from tiny area in W Himalayas; not definitely recorded since 1870s.

Tall grass, hill scrub; Endemic; perhaps extinct

39.7. Himalayan Quail
Ophrysia superciliosa

25cm   p. 129

Small and dark with red bill and legs, and pre- and postocular white spots. Male is slate-grey with black streaks, and black face and throat with white brow and cheek-spot. Female bright tawny-rufous with broad grey brow, cheek and throat, and heavy black streaks. In flight, browner flight feathers, and longish, broad tail.

Megapode Megapodiidae

Distinctive large dark galliform that incubates its eggs in a mound of rotting vegetation. Newly hatched chicks dig their way out and can fly right away.

Coastal forest; paler, darker, browner; Endemic

39.8. Nicobar Megapode
Megapodius nicobariensis

43cm   p. 118

Large, chunky and short-tailed with immense feet; dark brownish with grey head, reddish facial skin and pale bill.

PLATE 40. QUAILS AND BUTTONQUAILS (J. Schmitt)

Quail *Coturnix, Perdicula*

Tiny, very short-tailed galliformes, most cryptic in plumage and behaviour, and thus difficult to identify, especially the drabber females. Note especially face pattern.

40.1. Common Quail
Coturnix coturnix

20cm   p. 123

Broad buffy brow and throat-patch, and pale streaks above; see Japanese and female Rain. Male has blackish border to throat, dark whisker-mark and variable brown throat-bar, whitish shaft-streaks on buffy breast and black-and-buff flank streaks. Female has small cheek-mark and dark spots on breast.

40.2. Japanese Quail
Coturnix japonica

20cm   p. 123

Not always separable from Common apart from breeding male, which has rufous throat and face. Non-breeding male typically has heavier rufous streaks on flanks, and often a gorget of chestnut and blackish spots. Female identical to female Common, but often more boldly marked below.

40.3. Rain Quail
Coturnix coromandelica

18cm   p. 124

Small. Male has black breast-patch and streaks on flanks, and bolder head pattern than male Common. Female from female Common by finer spots on breast and less heavily streaked flanks.

40.4. Painted Bush-quail
Perdicula erythrorhyncha

16cm   p. 125

Red bill and legs, and rufous below with black spots on flanks. Male has black face, and white brow and throat. Female has rufous brow and throat.

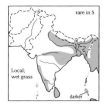

40.5. Blue-breasted Quail
Coturnix chinensis

14cm   p. 124

Male slate-blue with white lores, face patch and crescentic gorget, and chestnut central underparts. Female similar to female Common but smaller, with blackish bars on breast and more uniform upperparts.

40.6. Manipur Bush-quail
Perdicula manipurensis

20cm   p. 126

Dark with a white eye-patch; scaled above with fulvous spots below. Male has chestnut brow, cheek and throat. Female is pale below.

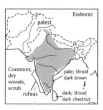

40.7. Jungle Bush-quail
Perdicula asiatica

17cm   p. 124

From Rock by chestnut brow and throat both outlined above with white, brown cheek-patch, coarsely black-blotched and pale shaft-streaked upperparts, and rufous under tail-coverts. Both species highly variable in depth of overall colour.

40.8. Rock Bush-quail
Perdicula argoondah

17cm   p. 124

From Jungle by mostly rufous face separated only by narrow line, and not bordered above by white, finely marked upperparts, and pale under tail-coverts. Varies in colour from very pale to bright rufous overall.

Buttonquails *Turnix*

Very small terrestrial birds of unclear affinities, found in open, often disturbed habitats; remarkably quail-like, but note the stout and usually pale bill. Females generally larger and more colourful, but plumages are complex and some not well understood. Note pattern of breast and upperparts.

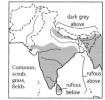

40.9. Barred Buttonquail
Turnix suscitator

15cm   p. 137

Rather dark with salt-and pepper head, black bars on breast and unmarked rufous belly. Female has black throat and centre of breast.

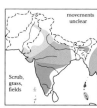

40.10. Yellow-legged Buttonquail
Turnix tanki

15cm   p. 137

Diagnostic yellow legs and round black spots on flanks and wing; dark crown may have pale median stripe or black-and-white stippling (as in Barred). Female usually has broad rufous hindcollar.

40.11. Small Buttonquail
Turnix sylvaticus

13cm   p. 137

Small with *Coturnix*-like upperparts, but with scaly rufous hindcollar, mostly buff wing; less distinctly spotted on wing than Yellow-legged, and whiter below.

PLATE 41. SPURFOWL AND JUNGLEFOWL (J. Schmitt)

Spurfowl *Galloperdix*

Endemic genus of mid-sized galliformes, seemingly intermediate between partridges and junglefowl; lack head and tail ornamentation and bright colours of latter but have rather long, somewhat arched tails and strong sexual dimorphism.

Junglefowl *Gallus*

Forest pheasants with rather short vaulted tails and elongated mane-like neck plumage. Males have red facial wattles and tall combs, and are spectacularly coloured with elongated back plumage and tail-coverts; males of Red and Grey assume eclipse plumage in summer with reduced wattles, no plumes, and cape replaced by short blackish-brown feathers. Immatures are more or less like females.

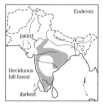

41.1. Red Spurfowl
Galloperdix spadicea

36cm   p. 128

Rufous with greyish head, red facial skin and legs and longish dark tail. Female is browner, entirely finely barred black. In flight has short, rounded wings with darker primaries and tail.

41.2. Painted Spurfowl
Galloperdix lunulata

32cm   p. 128

Dark with buffy belly, dark bill and legs, and no red facial skin. Male has blackish upperparts and breast spotted and barred with white (throat can appear whitish), with chestnut patches on mantle, wing, rump and flanks. Female is dark brown above, paler on breast and belly, with rufous brow and ear-coverts, and pale throat.

41.3. Ceylon Spurfowl
Galloperdix bicalcarata

34cm   p. 129

Red bill, facial skin and legs; chestnut upperparts with 'salt-and-pepper' head and black tail. Male has black-and-white streaks on head, underparts, mantle and wing-coverts. Female is solid chestnut, with head like male. Sympatric female Ceylon Junglefowl is larger, duller and browner with barred wings, and is marked black and white on breast and belly.

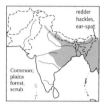

41.4. Red Junglefowl
Gallus gallus

male 66cm/female 43cm   p.132

Grey legs. Male has 'toothed' comb, orange cape, flight feathers and back plumes with white rump, and black underparts, tail and multiple tail-plumes. Female is dull brown with rufous head and breast, reddish facial skin, and black scales on golden-buff nape; lacks greyish head and red legs of Red Spurfowl. In flight, female shows less contrast above than female Grey.

41.5. Grey Junglefowl
Gallus sonneratii

male 70cm/female 46cm   p.132

Male is grey with heavy buff spots on cape, and black wings and tail with long, narrow central feathers, and yellowish legs. Female from female Red by pale buffy face, black-and-white-streaked underparts and yellowish legs. In flight, female shows more contrast between blackish primaries and tail and brown upperparts than female Red.

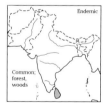

41.6. Ceylon Junglefowl
Gallus lafayetii

male 69cm/female 36cm   p.132

Red legs. Male is burnt-orange with black flight feathers and tail with central plumes. No eclipse. Female has black-barred flight feathers, and black scales on whitish underparts; as all subadult plumages, lacks greyish head of smaller female Ceylon Spurfowl.

PLATE 42. BLOOD PHEASANT, TRAGOPANS AND MONALS (H. Burn)

Blood Pheasant *Ithaginis*

Odd, small, partridge-like, short-tailed montane pheasant.

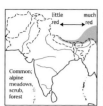

42.1. Blood Pheasant
Ithaginis cruentus

46cm   p.129

Small pheasant with short unmarked tail and reddish legs. Male pale grey with bold pale shaft-streaks, red throat and facial skin bordered by black, and pale green underparts. Colour of forehead (red/black) and amount of red on breast and vent geographically variable. Female brown with grey nape and rufous face and throat.

Tragopans *Tragopan*

Stocky, small-billed short-tailed pheasants of high Himalayas. For males, note head and belly pattern. For females, note eye-ring colour and coarseness of markings overall.

42.2. Satyr Tragopan
Tragopan satyra

male 68cm/female 59cm   p.130

Male has black-bordered white ocelli on red underparts, and no white on upper tail-coverts; facial skin darker than Temminck's, and back and rump grey-brown vs red. Female brownish-grey to rufescent, with bluish orbital skin and brighter rufous-brown wings and tail than other female tragopans; lacks white streaks below. Subadult male as female but head blacker, with some red on foreparts.

42.3. Temminck's Tragopan
Tragopan temminckii

64cm   p.130

Male has bright blue facial skin, red back and rump, pale upper tail-coverts, and red underparts with broad pale spots without black borders; see male Blyth's. Female greyish to rufescent with broad whitish streaks below; upperparts generally more heavily marked than other female tragopans. Subadult male as female, with a few male characters; weaker ocelli below than subadult Satyr.

42.4. Western Tragopan
Tragopan melanocephalus

71cm   p.130

Male is very dark with white ocelli, crimson hindneck-patch, orange foreneck, reddish facial skin and white uppertail-coverts with black tips. Female greyer and more finely marked than other fe-

male tragopans, without blue facial skin. Subadult male as female with darker head and reddish collar. In flight appears large and dark with broad, rounded tail.

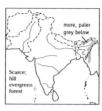

42.5. Blyth's Tragopan
Tragopan blythii

male 68cm/female 59cm   p.130

Male has diagnostic yellow facial skin, whitish belly, and rufous-and-white uppertail-coverts. Female uniformly darker, more olive than other female tragopans, with yellow eye-ring. Subadult male (not shown) as female but has blacker head with red on foreparts.

Monals *Lophophorus*

Stocky, large-billed, short-tailed pheasants of high Himalayas. For males, note crest type, rump and tail pattern. For females, note pattern of head and underparts.

42.6. Himalayan Monal
Lophophorus impejanus

72cm   p.131

Small blue eye-patch and dark bill. Male has peacock-like crest, very glossy upperparts with greenish upper mantle and small white back-patch, and rufous tail with no white tip. Female has short crest and white throat, long white streaks below, whitish band on uppertail-coverts; black and chestnut banded tail.

42.7. Sclater's Monal
Lophophorus sclateri

72cm   p.131

Uncrested with large blue eye-patch and pale bill. Male has bronzy shoulder and upper neck, and large white rump-patch; tail is dark chestnut with white tip in E Arunachal, entirely white in W Arunachal. Female has speckled head, weak whitish throat, blackish tail, and greyish underparts, without white streaks and bold white throat of female Himalayan. Subadult male as female with more blue on face, darker throat, coarser barring below, whiter rump, and more boldly banded tail.

PLATE 43. SHORTER-TAILED PHEASANTS AND SNOWCOCKS (H. Burn)

Eared-Pheasant *Crossoptilon*

Large, distinctively plumaged montane pheasant virtually lacking sexual dimorphism; even juvenile much like adults.

43.1. **Tibetan Eared-pheasant**
Crossoptilon harmani

72cm p. 133
Hypothetical; NE. Dark with paler rump and belly, white throat-patch extending to ear-tufts, red face and legs and heavy, vaulted black tail.

Peacock-Pheasant *Polyplectron*

Medium-sized lowland forest pheasant, rather dull coloured except for glossy ocelli; sexual dimorphism moderate.

43.2. **Grey Peacock-pheasant**
Polyplectron bicalcaratum

64/48cm p. 134
Grey with broadly rounded tail and whitish throat-patch. Male has short nasal crest and large blue ocelli on upperparts. Female and subadults are smaller, uncrested, and browner with less conspicuous ocelli.

Pheasants *Lophura, Pucrasia*

Medium-sized hill forest pheasants, mostly dark-plumaged with conspicuous crests and mid-length tails; strongly sexually dimorphic and highly geographically variable.

43.3. **Kaleej Pheasant**
Lophura leucomelanos

64cm p. 133
Dark with long, low crest, red face and arched, rooster-like tail. Male has black head, mantle, wings and tail; otherwise highly variable: grey below with white scales on rump in W and C Himalayas; black-rumped in E Nepal to W Bhutan; black below with white bars on rump in NE. White-crested only in W Himalayas. Female brown, heavily scaled pale, with pale head and crest, and uniform dark wings and tail.

43.4. **Koklass Pheasant**
Pucrasia macrolopha

61/53cm p. 131
Medium-sized with white neck-patch, pointed occipital crest and short broad tail. Male has black head with buffy crest and long black ear-tufts, chestnut underparts, and blackish upperparts trimmed with buffy-grey. Mostly buffy-grey above in W, blacker overall in C. Female dark brown with pale shaft-streaks, buff brow and white throat. Larger female Himalayan Monal lacks brow, and has blue eye-patch. In flight, stocky with short, rounded wings and broad, wedge-shaped tail.

Monal-partridge *Tetraophasis*

Medium-sized montane rather partridge-like galliform, strikingly patterned in earth tones and essentially lacking sexual dimorphism; juvenile differs.

43.5. **Széchenyi's Monal-partridge**
Tetraophasis szechenyii

46cm p. 119
Hypothetical; NE. Grey-brown with red orbital skin, buff throat and belly, grey breast, pale bars on scapulars and wings, and broad white tail-tip. In flight, grey with pale bars on inner wing and broad-based tapering grey tail with subterminal black band and white tips.

Snowcocks *Tetraogallus*

Large, bulky, partridge-like grey galliforms of high mountains, with distinctive white, chestnut, and black patterning; sexual dimorphism essentially lacking; juvenile differs.

43.6. **Himalayan Snowcock**
Tetraogallus himalayensis

68cm p. 119
Very large; grey with mostly white head and breast, black moustachial stripe and bars on throat, dark chestnut necksides, buff streaks on upperparts, and white vent. In flight note almost white primaries with dark tips, and dull rufous outertail feathers.

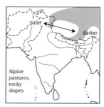

43.7. **Tibetan Snowcock**
Tetraogallus tibetanus

51cm p. 119
From larger Himalayan by grey head with white postocular stripe, white underparts with black stripes, and white (*vs* rufous) trim on wing. In flight has broad white trailing edge to secondaries, dark primaries and grey tail. Snow Partridge (Pl. 39) has narrower white trailing edge, and outertail feathers black tipped rufous.

Plate sponsored by Mr. Paul Mellon.

PLATE 44. LONG-TAILED PHEASANTS AND PEAFOWL (H. Burn)

Typical pheasants *Catreus, Syrmaticus, Phasianus*

Rather large pheasants with long narrow barred tails, usually bare red faces and grey legs. Strongly sexually dimorphic, males mostly with highly ornamented plumage, females cryptic.

44.1. **Cheer Pheasant**
Catreus wallichii

90–118cm/61–76cm　　　　　p. 133

Long pointed barred tail, recumbent dark crest and red facial skin. Male is grey with rufous rump and belly, and sandy tail with broad dark brown bands. Female is brown above with whitish bands on wings, and rufous below with white throat and blackish breast. In flight, barred tail and pale barring on wings distinctive.

44.2. **Mrs Hume's Pheasant**
Syrmaticus humiae

90/60cm　　　　　p. 134

Uncrested with pointed tail and red orbital skin. Male is very dark with bold white wing-bars, silvery-grey rump and long tail with dark bars. Female is warm brown, mottled blackish above and on breast, with whitish wing-bars and pale rufous below with white scales; shortish tail has chestnut sides and broad white tip. In flight, female from Mountain Bamboo-partridge (Pl. 39) by longer tail with white-tipped chestnut outer feathers, and lack of rufous in primaries.

44.3. **Common Pheasant**
Phasianus colchicus

80/60cm　　　　　p. 135

Black bars on long pointed brown tail. Male is chestnut with black head and neck, red facial skin, and large white wing-patch; paler and glossed purplish below in NW Afghanistan, darker and green-glossed in NC Afghanistan. Allopatric male Mrs Hume's is darker and grey-tailed, with two white wing-bars. Female is pale buff-brown with dark bars on crown and neck, and broader spots on mantle and flanks. Female Mrs Hume's is much more richly coloured, especially on tail, with red orbital skin.

Peafowl *Pavo*

Very large, long-legged and long-necked pheasants with upstanding crests, rufous primaries, and strong face markings; sexual dimorphism moderate. Breeding males have greatly elongated upper tail-coverts that form elaborate display fans.

44.4. **Indian Peafowl**
Pavo cristatus

male 110cm (2–2.5m w/train);
female 86cm　　　　　p. 134

Very large and long-necked, with upright crest, rufous primaries (concealed at rest), and white on cheek and brow. Male has blue neck, barred black-and-white mantle and long train with large blue-and-copper ocelli. Female lacks train; has green neck, white underparts and solid brown upperparts.

44.5. **Green Peafowl**
Pavo muticus

male 110cm (2–2.5m w/train);
female 86cm　　　　　p. 135

Hypothetical; NE. Darker than Indian with tall spike-like crest. Male has blue and yellow facial skin, bronzy-green neck, and lacks male Indian's white face-patches and black-and-white barred mantle. Female is dark-bellied and lacks train.

Plate sponsored by Mr. Perry Bass.

PLATE 45. CRANES (J. Anderton)

See Plate 46 for flight illustrations

Cranes *Grus*

Long-necked, long-legged wading birds with straight stout bills and elongated tertial plumes overhanging tail, forming 'bustle'. Neck outstretched in flight, unlike herons. Note head and neck pattern, plume colour, size.

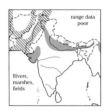

45.1. Common Crane
Grus grus

114cm p.131

Mid-sized grey crane with black face, nape and foreneck, small red crown-patch, broad white hindneck-stripe, and black spots on wing-coverts. Juv. has brownish head with ghost of adult pattern.

45.2. Black-necked Crane
Grus nigricollis

139cm p.140

Large and pale with solid black head and neck, white postocular spot, red crown and blackish bustle. Juv. is washed brown above, has black areas of adult dark greyish.

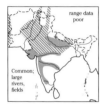

45.3. Demoiselle Crane
Grus virgo

95cm p.139

Small; from Common by white head-plumes, grey crown, and black of neck extending to breast-plumes; bustle longer and more tapered. Juv. largely grey and lacks plumes.

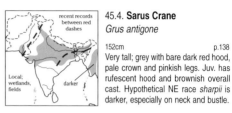

45.4. Sarus Crane
Grus antigone

152cm p.138

Very tall; grey with bare dark red hood, pale crown and pinkish legs. Juv. has rufescent hood and brownish overall cast. Hypothetical NE race *sharpii* is darker, especially on neck and bustle.

45.5. Siberian Crane
Grus leucogeranus

135cm p.138

Tall and white with red face and bill, and black primaries. Juv. is heavily washed rufous; lacks grey

45.6. Hooded Crane
Grus monacha   97cm p.139

Hypothetical; NE. Small and dark grey with white hood, black face and small red crown-patch.

PLATE 46. CRANES AND BUSTARDS IN FLIGHT (J. Anderton)

See Plate 45 for non-flight crane illustrations, Plate 47 for bustards

46.1. Common Crane
Grus grus p.139

Black flight feathers (and carpal patch from above) and grey tail; see Demoiselle.

46.2. Black-necked Crane
Grus nigricollis p.140

From Common by black tail, little white on head.

46.3. Sarus Crane
Grus antigone p.138

Pale with black primaries; pattern like pure white Siberian.

46.4. Demoiselle Crane
Grus virgo p. 139

From Common by black breast, and weaker contrast between pale coverts and blackish primaries. From above, no black carpal patch.

46.5. Siberian Crane
Grus leucogeranus p.138

All white with black primaries.

46.6. Hooded Crane
Grus monacha p.139

Entirely dark with white head and neck.

46.7. Little Bustard
Tetrax tetrax p.148

In both sexes, white wings with black tips, black carpal patch.

46.8. Great Bustard
Otis tarda p.148

In both sexes, white wings with broad black trailing edge; from above, note rufous coverts and black carpal-crescent.

46.9. Lesser Florican
Sypheotides indicus p.149

Breeding male has white coverts on upperwing. Female plainer-winged with blackish carpal patch and axillaries.

46.10. Bengal Florican
Houbaropsis bengalensis p.149

Breeding male has entirely white wing with fine black tips on primaries. Female from female Lesser by pale wing-lining and darker flight feathers.

46.11. Houbara Bustard
Chlamydotis undulata p.148

Black flight feathers with large white patch on base of primaries. From above, buff coverts and black carpal patch; from below, white wing-lining.

46.12. Great Indian Bustard
Ardeotis nigriceps p.148

Black flight feathers. From above, little white; from below, white wing-lining.

PLATE 47. BUSTARDS (J. Anderton)

See Plate 46 for flight illustrations

Bustards Otididae

Very large to medium-sized cryptically plumaged terrestrial birds with rather short stout bills and long legs and neck; males have elaborate displays. All species rare and local in region. Note size, head and upperpart pattern.

Floricans Houbaropsis, Sypheotides

Breeding males have glossy black head and underparts, and much white in wing. Females and non-breeding males are mottled brown with striped heads and largely pale wings; females larger than males.

Rare, local, grasslands

now within red dashes

47.1. Bengal Florican
Houbaropsis bengalensis

66cm   p.149

Breeding male larger and thicker-necked than breeding male Lesser, with no white on neck; darker and more coarsely streaked above. Nonbreeding male like female but lower underparts black. Female is similar to female Lesser but duller buff with diffuse facial pattern and plainer neck.

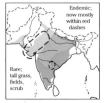

Rare; tall grass, fields, scrub

Endemic; now mostly within red dashes

47.2. Lesser Florican
Sypheotides indicus

48cm   p.149

Small, slim-necked and 'bony-headed'. Breeding male from breeding male Bengal by long black head-plumes, and white throat and upper back. Female much larger; brighter golden-buff than female Bengal, with sharper dark streaks on face and throat, blotches on breast and large black triangles on mantle. See stone-curlews (Pl. 51).

Bustards Tetrax, Chlamydotis, Otis, Ardeotis

Diverse; in most species males show little seasonal plumage change, and are much larger than females.

Rare; plains grass, crops

47.3. Little Bustard
Tetrax tetrax

46cm   p.148

Small and short-billed; tawny with white belly. Breeding male has grey head and black-and-white collar.

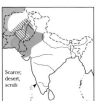

Scarce; desert, scrub

47.4. Houbara Bustard
Chlamydotis undulata

65cm   p.148

Medium-sized and long-bodied; pale sandy with barred upperparts, grey foreneck and white belly. Black stripe on neck normally mostly hidden.

Vagrant; open steppe, fields

47.5. Great Bustard
Otis tarda

102cm   p.148

Very large and heavy, with grey head, black bars on rufous upperparts, and white belly. Breeding male has bushy white 'moustache' and rufous breast, both lost in non-breeding; female much smaller, with thinner neck.

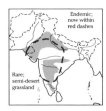

Rare; semi-desert grassland

Endemic; now within red dashes

47.6. Great Indian Bustard
Ardeotis nigriceps

110cm   p.148

Huge, especially male. Brown above with whitish or grey head and neck, and black crown and short crest. Female is smaller and duller with weaker breast-band. In display, male fluffs and droops white neck-plumes, cocks tail.

Plate sponsored by Mr. Perry Bass.

PLATE 48. BARRED AND RUFOUS RAILS (H. Burn)

Rails and Crakes Rallidae

Rather to very small, most with barred flanks and/or generally rufous plumage. Sexual dimorphism weak to moderate.

48.1. Slaty-breasted Rail
Rallus striatus

27cm p.141

Fine white bars on blackish upperparts diagnostic; blue-grey face and breast, rufous crown and hindneck, longish red bill. Juv. paler, olive-brown. Larger, darker in Andamans and Nicobars, where juv. very dark, with narrower barring.

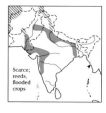

48.6. Spotted Crake
Porzana porzana

23cm p.143

Larger and bigger-billed than 'Eastern' Baillon's; paler with spotted foreparts, unbarred buff vent, red bill-base and brown eye. Juv. duller and more heavily spotted. In flight wing has broad white leading edge.

48.2. European Water Rail
Rallus aquaticus

28cm p.141

Streaked upperparts with slate-grey face and underparts, white vent, long reddish bill and pinkish legs. Juv. long-billed, mottled dusky brown face and underparts with paler brown throat and centre of underparts.

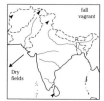

48.7. Corn Crake
Crex crex

25cm p.142

Chunky; pale grey-and-buff with rufous-barred flanks and short pinkish bill. In flight chestnut wings striking.

48.3. Eastern Water Rail
Rallus indicus

28cm p.141

From paler Western by brown eye-stripe, browner breast and upper flanks, and black-and-white-barred vent. Larger and bigger-billed than Baillon's and Little Crakes. Juv. has browner face and underparts.

48.8. Andaman Crake
Rallina canningi

34cm p.140

Large and entirely chestnut with bold black-and-white bands on belly, green bill, red eye and olive legs. Juv. is duller.

48.9. Red-legged Crake
Rallina fasciata 23cm p.140

Hypothetical; NE. Bright chestnut head with dark bill, black-and-white bars on wings as well as belly, and red eye and legs.

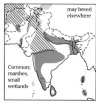

48.4. 'Eastern' Baillon's Crake
Porzana pusilla

19cm p.143

Tiny; marbled white above, with strongly barred flanks, black marks on crown, olive legs, and no red on bill. Female duller but less so than in Little. Short primary projection; in flight wing has whitish leading edge. Juv. is buffy with extensive white spots on wing-coverts, bars on flanks.

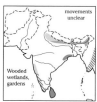

48.10. Slaty-legged Crake
Rallina eurizonoides

25cm p.140

Chestnut head with dark bill and legs, black-and-white barred underparts; bill and legs dark. Juv. has dull brown replacing rufous.

48.5. Little Crake
Porzana parva

20cm p.143

Tiny, usually with green legs, red eyes. From 'Eastern' Baillon's by red base of greenish bill, weaker whitish bars below, and longer primary projection. Male is grey on face and underparts; female and juv. are buff below.

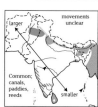

48.11. Ruddy-breasted Crake
Porzana fusca

22cm p.144

From Slaty-legged by red legs, and weak barring below restricted belly, legs red. Juv. is brown overall, with weakly barred foreparts.

PLATE 49. DUSKY RAILS, GALLINULES, COOT AND FINFOOT (H. Burn)

Dusky crakes and Waterhen *Porzana* (part) and *Amaurornis*

Dark rallids with little or no rufous, and lacking flank banding. Sexual dimorphism minimal.

49.1. Brown Crake
Porzana akool

28cm p.142
Rather large; olive-brown with grey face and breast, whitish throat, yellowish-green bill, and dark red legs.

49.3. White-breasted Waterhen
Amaurornis phoenicurus

32cm p.142
Dark with white face, throat and belly, and yellow bill and legs; whole crown often white in Nicobars. Juv. is greyish below.

49.2. Black-tailed Crake
Porzana bicolor

22cm p.142
Slaty with dark chestnut mantle and wings, black tail and red legs. Only entirely slaty-headed rail.

Watercock, Swamphen, Gallinule and Coot *Gallicrex, Porphyrio, Gallinula, Fulica*

Diverse dark-plumaged rallids that may recall galliformes or ducks. Sexual dimorphism minimal to strong.

49.4. Watercock
Gallicrex cinerea

43cm/36cm p.144
Large and lanky with heavy legs and yellowish bill. Breeding male is blackish with red frontal shield and legs. Non-breeding male and smaller female are buffy with buff-scaled dark brown upperparts and yellowish legs; variable.

49.6. Common Moorhen
Gallinula chloropus

42cm p.146
Rather plump; blackish with yellow-tipped red bill and shield, white flank-stripe and vent, and olive legs. Bill is brownish-red in non-breeder. Juv. dull and brownish with pale face and belly, dark bill.

49.5. Purple Swamphen
Porphyrio [*porphyrio*] *poliocephalus*

43cm p.144
Huge; purplish-blue with pale head, massive red bill and legs, and white vent. Head whitish in W. Juv. has dark bill.

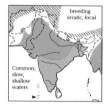

49.7. Eurasian Coot
Fulica atra

42cm p.147
Blackish-slate with white bill and frontal shield and greenish legs. Juv. is duller; more uniform than young Common Moorhen. Usually swimming.

Finfoot Heliornithidae

Strange, long-bodied, somewhat cormorant-like waterbird.

49.8. Masked Finfoot
Heliopais personatus

56cm p.147
Long-bodied and buff-brown with black face and crown-stripe, stout yellow bill and longish tail; male has black throat, white in female. Swims; can dive.

PLATE 50. PAINTED-SNIPE, RECURVIROSTRIDS, CRAB-PLOVER, OYSTERCATCHER AND JACANAS (J. Alderfer; flights of 1–6 by J. Schmitt)

Painted-snipe Rostratulidae

Snipe-like, crepuscular marsh bird; female much brighter than male.

50.1. Greater Painted-snipe
Rostratula benghalensis 25cm Pl. 58, p.151

Map on Pl. 58. Snipe-like with droop-tipped bill, white postocular stripe, and golden mantle-stripes. Female has chestnut head and dark back; male is duller with gold-spotted upperparts.

Ibisbill, Crab-plover, Avocet and Stilt
Recurvirostridae, Dromadidae

Diverse, strikingly plumaged, long-legged and long-necked shorebirds of open waters. Sexual dimorphism and seasonal variation weak.

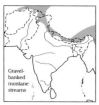

50.2. Ibisbill
Ibidorhyncha struthersii

42cm p.178
Grey with black face and breast-band, and curved red bill. In flight, white patch at base of primaries, and underwing white.

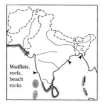

50.3. Crab-plover
Dromas ardeola

41cm p.181
White with large black bill, black mantle and flight feathers. Juv. is grey in place of black.

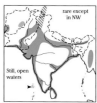

50.4. Pied Avocet
Recurvirostra avosetta

46cm p.180
Pied, with long slender upturned bill, long bluish legs.

50.5. Black-winged Stilt
Himantopus himantopus

25cm p.179
White with black mantle and wings, extremely long red legs, and long thin bill. Non-breeder has variable dark on hindneck and/or crown. See text for hypothetical Australian *H.* [*h.*] *leucocephalus*.

Oystercatcher Haematopodidae

Large, chunky, strikingly marked coastally-wintering shorebird. Sexual dimorphism in dimensions only; some age and seasonal variation.

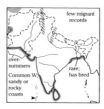

50.6. Eurasian Oystercatcher
Haematopus ostralegus

42cm p.151
Black, with white belly, heavy red bill, red eye, eye-ring and legs. Non-breeder has white chinstrap; juv. dusky overall. In flight, white rump, wing-stripe and base of tail.

Jacanas Jacanidae

Rail-like shorebirds with extremely long toes and claws used for walking on floating vegetation. Note head pattern, wing colour, tail length.

50.7. Bronze-winged Jacana
Metopidius indicus

30cm p.150
Dark and chunky, with white postocular stripe, yellow bill with red shield, and short chestnut tail. Juv. is whitish below with rufous neck; lacks dark breast-band/neck-stripe of juv. Pheasant-tailed; in flight wings all dark.

50.8. Pheasant-tailed Jacana
Hydrophasianus chirurgus

31cm w/o streamers p.150
White throat in all plumages. Breeder is dark with sweeping black tail, golden nape and white wings; non-breeder has dark breast-band and long golden brow.

Plate sponsored by Mr. Perry Bass.

PLATE 51. COURSERS, PRATINCOLES AND THICK-KNEES
(J. Alderfer; flights of 7, 8, 10 by J. Schmitt)

Pratincoles *Glareola*

Gregarious, short-legged, short-necked, short-billed, long-winged relatives of coursers; crepuscular Collared and Oriental share rufous underwing-linings. Hawk insects on wing. Note tail length and shape, wing pattern.

rare in S
Open ground near water

51.1. Collared Pratincole
Glareola pratincola

23cm   p.184
Paler than Oriental; in flight, longer, more forked tail and white trailing edge to secondaries.

erratic; breeding, movements unclear
Open ground near water

51.2. Oriental Pratincole
Glareola maldivarum

24cm   p.185
Darker than Collared; at rest, tail far short of wing-tips. May have light rufous upper belly.

Large sandy riverbanks, marshes, estuaries

51.3. Small Pratincole
Glareola lactea

17cm   p.185
Delicate, very pale plain sandy-grey with black primaries; in flight, black wing-lining and tail-patch and mostly white secondaries.

Coursers *Rhinoptilus, Cursorius*

Slim and long-legged, fast-running shorebirds of dry inland habitats, with shortish bills and conspicuous white brows. Juveniles like adults but mottled. Note breast, belly and wing pattern in flight.

Endemic
Rare; thin thorn scrub

51.4. Jerdon's Courser
Rhinoptilus bitorquatus

27cm   p.183
Brown with white throat and belly, straight bicoloured bill, large eye, and banded breast. In flight, black flight feathers with white-tip and wing-lining.

Barren desert, near villages

51.5. Cream-coloured Courser
Cursorius cursor

23cm   p.184
Bright buff-brown, with black postocular line, and grey hindcrown. In flight, black wing-lining.

Endemic
Fields, open scrub, gullies

51.6. Indian Courser
Cursorius coromandelicus

26cm   p.184
Darker than Cream-coloured, with chestnut crown, black stripe before eye, rufous neck and breast, and blackish belly with white vent. In flight, wing-lining brownish.

Thick-knees *Burhinus, Esacus*

Brownish shorebirds with large yellow eyes, heavy yellow legs, thick yellow-based black bills, dark cheek-patches, and striking wing patterns. Note size and shape of bill.

Reefs, beaches

51.7. Beach Thick-knee
Esacus magnirostris

55cm   p.183
From Great by deeper, straighter bill, and dark lores and forehead.

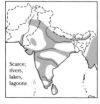

Scarce; rivers, lakes, lagoons

51.8. Great Thick-knee
Esacus recurvirostris

51cm   p.183
Pale and unstreaked with huge, slightly recurved bill. In flight, upperwing more extensively black than Beach.

NW limits unclear
larger, paler
smaller, darker
Dry scrub, groves, gardens

51.9. Indian Stone-curlew
Burhinus indicus

<41cm   p.181
Mid-sized, with streaked buffy plumage. Smaller and darker than Eurasian, with little yellow on larger bill, longer legs and paler panel on median wing-coverts.

SE limits, movements unclear
Scarce; desert

51.10. Eurasian Stone-curlew
Burhinus oedicnemus

>41cm   p.181
Larger and paler than Indian; relatively smaller bill has more yellow at base.

PLATE 52. LAPWINGS (J. Alderfer)

Lapwings *Vanellus*

Large, mostly inland plovers with broad wings; strikingly marked, especially in flight, with black primaries and, in most, bold white panels. Note head, wing and tail patterns, and bill and leg colour.

52.1. **Red-wattled Lapwing**
Vanellus indicus

33cm   p.158

Black crown, face and breast with red facial skin and bill; white cheek is enclosed by black in NE race *atronuchalis*. Juv. has whitish throat. In flight, wing and tail pattern like Yellow-wattled.

52.2. **Yellow-wattled Lapwing**
Vanellus malabaricus

27cm   p.158

White stripe from eye to nape, dark crown (black in breeder) above white eye-stripe and yellow facial wattles. Juv. is mottled above with reduced wattles. In flight, broad white wing-bar and narrow white rump-band.

52.3. **River Lapwing**
Vanellus duvaucelii

31cm   p.158

Black crown, face, throat, bill and legs. In flight, long white wing-bar behind small black wrist-patch, mostly black tail, and small black belly-patch on white underparts.

52.4. **Grey-headed Lapwing**
Vanellus cinereus

37cm   p.158

Plain grey head with yellow bill and dark breast-band. Juv. is browner-headed with mottled upperparts. In flight, secondaries entirely white.

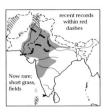

52.5. **Sociable Plover**
Vanellus gregarius

33cm   p.159

Dark cap and eyeline with whitish brow; dark legs. Breeder has ochraceous face and dark belly; juv. is scaly above and streaked on head and breast. In flight recalls Grey-headed and White-tailed, but white on secondaries does not reach leading edge. In NW, see text (p. 157) for vagrant Eurasian Dotterel *Eudromias morinellus*.

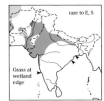

52.6. **White-tailed Lapwing**
Vanellus leucurus

28cm   p.159

Plain with gangly yellow legs. Juv. has black blotches on upperparts. In flight, white secondaries, carpal patch and diagnostic unmarked tail.

52.7. **Northern Lapwing**
Vanellus vanellus

31cm   p.157

Dark above and on broad breast-band, with dark wispy crest and face-patch on white head, and white below with rufous vent; dark legs and bill. Breeding male has black throat. Juv. browner and short-crested. In flight, wings very rounded and dark with white spots on tips, and mostly dark tail.

Plate sponsored by Mrs. Evelyn Bartlett.

PLATE 53. PLOVERS (J. Alderfer; flights and breeding males of 6–7 by J. Schmitt)

Plovers *Pluvialis, Charadrius*

Pluvialis plovers are marbled above, black below, with white side-stripes in breeding plumage. In moult, show irregular black patches below. *Charadrius* plovers are brown above and whitish below, with striking markings on crown and breast, typically weaker or lost in breeding plumage.

53.1. Grey Plover
Pluvialis squatarola

31cm　　　　　　　　　　p.152
Larger and greyer than golden-plovers, with heavier bill. Non-breeder very pale. In flight, black axillaries, narrow white wing-stripe, and white rump. Juv. has buffy tone above.

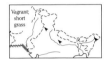

53.2. European Golden Plover
Pluvialis apricaria

27cm　　　　　　　　　　p.152
Chunkier than Pacific, with white underwing; non-breeder has vague supercilium.

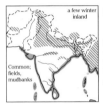

53.3. Pacific Golden Plover
Pluvialis fulva

24cm　　　　　　　　　　p.152
Slimmer and longer-legged than European, with heavier bill and, in flight, grey underwing. Non-breeder has gold-tinged pale supercilium.

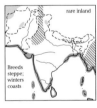

53.4. Greater Sand Plover
Charadrius leschenaultii

22cm　　　　　　　　　　p.155
This and Lesser lack white collar of smaller plovers. Greater is larger and usually paler than Lesser, with larger bill and longer, paler legs. Breeding male has white forehead spot and narrower rufous breast-band.

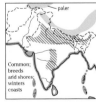

53.5. Lesser Sand Plover
Charadrius mongolus

19cm　　　　　　　　　　p.155
Smaller than Greater, with rounder head and shorter darker legs; bill usually no longer than distance from base of bill to rear of eye.

53.6. Oriental Plover
Charadrius veredus　　　　24cm　p.155

Vagrant; Sri Lanka, Andamans. From smaller Caspian by longer neck and legs (usually pinker). In flight, dark underwing, no white in primaries. Breeder (not shown) has plainer whiter head, and broader black breast-band.

53.7. Caspian Plover
Charadrius asiaticus

19cm　　　　　　　　　　p.155
Vagrant, W and S. Slim with broad pale eye-stripe. In flight, white underwing and white patch at base of inner primaries. Breeder (not shown) has chestnut breast bordered black below.

53.8. Common Ringed Plover
Charadrius hiaticula

19cm　　　　　　　　　　p.153
Thick bicoloured bill and bright yellowish-orange legs; in flight, strong white wing-stripe.

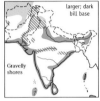

53.9. Little Ringed Plover
Charadrius dubius curonicus

17cm　　　　　　　　　　p.154
Small with a fine black bill, yellow eye-ring and dull legs. In flight (not shown), no wing-stripe. Juv. has buffy-fringed upperparts.

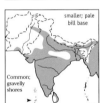

53.10. *Charadrius dubius jerdoni*

In most of region, *jerdoni* smaller, with more pale on bill, and no distinct non-breeding plumage.

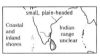

53.11. Kentish Plover
Charadrius alexandrinus seebohmi

17cm　　　　　　　　　　p.154
In S Indian and Sri Lanka breeding *seebohmi*, rufous cap and loral stripe lacking.

53.12. *Charadrius a. alexandrinus*

Dainty with black legs and bill, dark patch on breast-side. In flight, narrow white wing-stripe. In most of region, breeding nominate has rufous cap.

53.13. Long-billed Plover
Charadrius placidus

23cm　　　　　　　　　　p.153
Lanky, with longish dark bill, brown cheek (even in breeder), narrow complete breast-band, and long-tail. In flight, minimal white wing-stripe.

PLATE 54. CURLEWS, GODWITS AND DOWITCHERS (J. Alderfer)

Curlews *Numenius*

Large mottled shorebirds with long decurved bills. Note bill length, head-stripes.

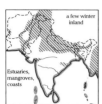

54.1. Eurasian Curlew
Numenius arquata

58cm   p.165
Large, uniformly streaked with white rump and belly; in flight, whitish underwing. See text for hypothetical Slender-billed *N. tenuirostris* (p. 167).

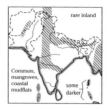

54.3. Whimbrel
Numenius phaeopus

43cm   p.164
Smaller and shorter-billed than Eurasian Curlew with striped crown, barred underwing. Vagrant E race *variegatus* has variably darker rump.

54.2. Far Eastern Curlew
Numenius madagascariensis   63cm   p.165

Hypothetical; Afghanistan, Bangladesh. From Eurasian by longer bill (in adult), dark rump and buff vent. In flight, dark (barred) underwing.

Godwits *Limosa*

Large shorebirds with long, straight or slightly recurved, pale-based bills. Note tail pattern, leg length.

54.4. 'Eastern' Black-tailed Godwit
Limosa [limosa] melanuroides

<46cm   p.163
Smaller and darker than 'Western', with darker forewing and smaller wing-stripe.

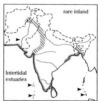

54.6. Bar-tailed Godwit
Limosa lapponica

39cm   p.164
Scaled above with shorter legs than congeners, and slightly upturned bill. In flight, plain wing and barred tail, white (or grey in E) rump and underwing. Breeder has chestnut head and underparts.

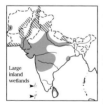

54.5. 'Western' Black-tailed Godwit
Limosa limosa

>46cm   p.163
White rump, wing-stripe and underwings, and black tail. Non-breeder plain grey above with white brow. Imm. marbled above, buffy on breast. Breeder has chestnut head and neck.

Dowitchers *Limnodromus*

Rather large, very long-billed shorebirds recalling godwits but stockier. Rare in region.

54.7. Asian Dowitcher
Limnodromus semipalmatus

34cm   p.172
From larger Bar-tailed Godwit by shorter neck and deep-based straight black bill, dark bars on flanks and on white rump. Non-breeder more scaled above, greyer below than godwits. Breeder has chestnut head and underparts.

54.8. Long-billed Dowitcher
Limnodromus scolopaceus   29cm   p.172

Vagrant; W. From Asian by less heavy bill with pale base. Non-breeder is plain grey overall with white belly and dark grey breast. Breeder is rusty-orange below with spots on breast, bars on flanks and vent. In flight more heavily barred rump and tail than Asian; underwings darker with more heavily marked axillaries and wing-lining. Juv. is pale tawny with rufous-edged dark mantle, grey wing-coverts and pale even edges on tertials.

PLATE 55. *TRINGINE* SANDPIPERS (J. Alderfer)

Tringine Sandpipers Xenus, Tringa, Actitis, Heteroscelus

Rather small to rather large, elegant shorebirds with more or less straight bills, and usually lacking wing-stripe. *Tringa* species are long-legged; other genera have shorter legs. Seasonal and age-related plumage changes minor to marked, sexual dimorphism minimal. Note bill shape, leg colour; in flight, pattern of rump and underside of wing.

Coastal lagoons, flats, reefs

55.1. Terek Sandpiper
Xenus cinereus

24cm p.169
Uptilted bill, short yellow legs; greyish with dark shoulder, whitish below. Breeder has black scapulars. In flight, white trailing edge to secondaries.

Rare; coastal mudflats

55.2. Spotted Greenshank
Tringa guttifer

33cm p.168
From Common Greenshank by shorter yellow legs and heavier bill with paler basal half. Non-breeder paler and less streaked. In flight, white underwing, unbarred tail.

Common; wetlands

55.3. Common Greenshank
Tringa nebularia

28cm p.168
Long green legs and long, slightly up-turned bill. Non-breeder streaked on breast and head. Breeder darker, more heavily streaked. In flight, dark wings with greyish underwing, white wedge on back, and barred tail with trailing legs.

Common; fresh wetlands

55.4. Wood Sandpiper
Tringa glareola

21cm p.169
From Green by stronger brow, more marbled upperparts and yellower legs; in flight by pale underwing, finer bars on tail. Longish legs and neck recall larger Common Redshank.

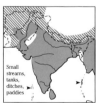
Small streams, tanks, ditches, paddies

55.5. Green Sandpiper
Tringa ochropus

24cm p.169
Mid-length bill and legs; darker and plainer above than Wood with weaker white eye-ring and brow, dark streaks on breast and sharply defined white belly. In flight, dark wings above and below, white rump and tail with broad dark bars.

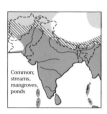
Common; streams, mangroves, ponds

55.6. Common Sandpiper
Actitis hypoleucos

36cm p.170
Shorter-legged and paler above than Green, with pale brown breast-sides. Constantly bobs hindquarters. In flight, broad white wing-stripe; fluttering wings alternate with brief glides.

Rare; coastal mudflats, fishponds

55.7. Grey-tailed Tattler
Heteroscelus brevipes

25cm p.170
Slate-grey above, on tail and breast, with white brow and shortish yellow legs. Breeder (not shown) has dark grey bars below. See text for Wandering Tattler.

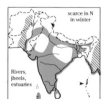
scarce in N in winter

55.8. Marsh Sandpiper
Tringa stagnatilis

25cm p.169
Long greenish legs, and small head with long very fine straight bill. Non-breeder very pale; darker and more streaked in breeding. In flight, dark wings, white rump and (mostly) tail.

Rivers, jheels, estuaries

55.9. Common Redshank
Tringa totanus

28cm p.168
Bright red legs and base of straight bill; grey-brown above with weak brow. In flight, broad white wedge on rear wing, and white back. Breeder and juv. browner, streaked.

Alpine breeder; winters mostly coasts

rare in S

55.10 Spotted Redshank
Tringa erythropus

33cm p.167
Greyer above than Common Redshank with longer bill and legs, stronger brow. In flight, no white in wing. Breeder sooty-black, finely spotted white above.

Scarce; fresh, brackish wetlands

PLATE 56. TURNSTONE AND LARGER CALIDRINES (J. Alderfer)

Turnstone *Arenaria*

Odd, chunky, strongly marked coastally-wintering shorebird. Major seasonal change but little age or sexual variation.

Larger *Calidris* sandpipers

Like stints (see Plate 57) but larger and somewhat more diverse. Seasonal, sexual and age variation minimal to strong.

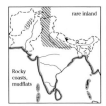

56.1. Ruddy Turnstone
Arenaria interpres

22cm p.170

Stocky with short stout black bill, orange-red legs, blackish breast and white belly. Breeder rufous above with harlequin facial pattern; non-breeder has dark brown head. In flight, double white wing-bars, white rump and tail with black tail-band.

56.5. Great Knot
Calidris tenuirostris

27cm p.173

Large; longish tapered bill, streaked crown and breast, spots on flanks, dark legs, whitish uppertail-coverts. Breeder is black-and-chestnut above, heavily spotted and scaled with black below.

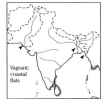

56.6. Red Knot
Calidris canutus

24cm p.173

Non-breeder from non-breeding Great by shorter bill, greyish rump, paler breast and tail, and pale greenish legs; imm. more scaled above. Breeder (not shown) bright rufous below.

Miscellaneous calidrine sandpipers
Limicola, Tryngites, Philomachus

Diverse rather small shorebirds clearly related to stints.

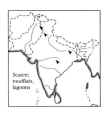

56.2. Broad-billed Sandpiper
Limicola falcinellus

17cm p.176

Stint-like, with longer, slightly decurved bill and diagnostic split pale brow. Breeder and imm. have clearer double brow, pale scapular- and mantle-lines, and streaked breast. In flight (not shown), narrow wing-bar and dark leading edge, white sides of rump.

56.3. Buff-breasted Sandpiper
Tryngites subruficollis 19cm p.176

Vagrant; Goa, SL. Buff with short thin bill and yellow legs; shorter-necked, shorter-legged than Reeve.

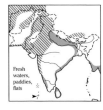

56.4. Ruff (male) and Reeve (female)
Philomachus pugnax

31/25cm p.178

Deep body with small head, slightly drooped bill, and long pale legs; male much larger than female. Non-breeder plain with buff-scaled dark upperparts. Breeding male (not shown) has long variegated head-plumes; breeding female blackish with pale scales. In flight, narrow wing-bar, whitish underwing, and white oval sides to rump.

56.7. Sharp-tailed Sandpiper
Calidris acuminata 19cm p.175

Vagrant; NW, SL. Ruff-like posture with fine, slightly decurved bill, dark rufescent cap, whitish brow and eye-ring. Non-breeder is mottled above, with fine streaks on greyish breast-band. Juv. has rufous cap, chestnut-and-white fringes on mantle, rufescent breast. Breeder similar with dark chevrons below. In flight like Ruff.

56.8. Pectoral Sandpiper
Calidris melanotos 21cm p.175

Vagrant; NW. From Sharp-tailed by sharp border between brown-streaked breast and white belly, less marked flanks and especially vent, and less rufous cap on weakly patterned head. Male much larger than female.

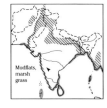

56.9. Curlew Sandpiper
Calidris ferruginea

21cm p.175

Non-breeder from non-breeding Dunlin by slightly longer bill, neck and legs, with stronger brow, whiter underparts. Imm. scaly above with buffish breast. In flight, white rump. Breeder scaly chestnut and black below. See text for hypothetical White-rumped *C. fuscicollis*.

56.10. Dunlin
Calidris alpina

20cm p.175

Hunched, with drooped bill-tip and black legs. Non-breeder plain grey above with white fringes, lightly streaked greyish breast; from stints by longer bill and size. Imm. has blackish spots on belly. In flight, clear white wing-bar and sides of rump. Breeder has black belly-patch.

PLATE 57. STINTS AND PHALAROPES (J. Alderfer)

Stints and other small *Calidris* sandpipers

Tiny, gregarious short-billed wintering waders, often hard to identify. Major seasonal and age but minimal sexual variation in plumage.

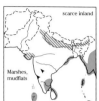

57.1. Long-toed Stint
Calidris subminuta

14cm p.174

Elongate, upright with streaked breast (especially in breeder) and yellowish legs; indistinct wing-bar in flight. Non-breeder is grey with whitish brow; darker and more patterned above than non-breeding Temminck's or Rufous-necked. Imm. has bolder white lines on mantle. Breeder has rufous crown and edges to black scapulars.

57.4. Little Stint
Calidris minuta

13cm p.174

Non-breeder almost identical to non-breeding Rufous-necked: longer-legged, shorter-winged, more hunched, with slightly longer and decurved bill, more extensive streaks on breast. Imm. has prominent white mantle-lines and brow with rufous crown. Breeder less rufous than breeding Rufous-necked.

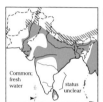

57.2. Temminck's Stint
Calidris temminckii

14cm p.174

Non-breeder is plain with little or no brow, grey-brown breast-band and yellowish legs. In flight, broad white wing-bar and white outertail. Breeder is browner and more mottled above.

57.5. Spoon-billed Sandpiper
Eurynorhynchus pygmaeus

15cm p.176

Swollen bill with spatulate tip; black legs. Non-breeder white on forehead and underparts. Imm. has darker brown and boldly patterned upperparts. Breeder like breeding Rufous-necked but upperparts with weaker golden fringes.

57.3. Rufous-necked Stint
Calidris ruficollis

14.5cm p.174

Non-breeder very like Little; greyer above than non-breeding Long-toed and Temminck's and almost unmarked white below with black legs. Imm. may have faint white mantle-lines; from imm. Little by less distinct brow behind eye. Breeder has pale rufous face, neck and breast, and mottled upperparts.

57.6. Sanderling
Calidris alba

19cm p.175

Stocky, with shortish black bill and legs. Non-breeder very pale with dark shoulder. Imm. has black centres to upperparts. Breeder has rufous breast, and rufous and black markings above. In flight, bold white wing-bar.

Phalaropes Phalaropodidae

Rather small, delicate-looking shorebirds that winter at sea. Marked seasonal and age variations in plumage; polyandrous breeding females much brighter in plumage than males.

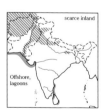

57.7. Red-necked Phalarope
Phalaropus lobatus

19cm p.180

Needle-like black bill. Non-breeder pale grey with black eye-stripe and hindcrown, white head and underparts. Breeder is dark with rufous on neck, and broad buff stripes above. In flight, narrow white wing-bar and sides of rump. Swims.

57.8. Red Phalarope
Phalaropus fulicarius

20cm p.180

Vagrant; NW and SE (not illustrated). Like Red-necked but bill thicker; bird bigger overall. Non-breeder paler above; breeder rufous below..

PLATE 58. PAINTED-SNIPE, SNIPES AND WOODCOCK (J. Schmitt)

Snipes *Gallinago, Lymnocryptes* and **woodcock** *Scolopax*

Long-billed, cryptically plumaged waders that inhabit marshes and rivers, and in some cases woodland streams; generally secretive and very difficult to identify. Painted-snipe looks similar but is unrelated. Note details of facial pattern, pattern of upperparts and belly, and, in flight, shape of wings and presence of white trailing edge, projection of feet beyond tail.

58.1. Greater Painted-snipe
Rostratula benghalensis

25cm   Pl. 50, p.151
Droopy bill-tip, white eye-ring and postocular stripe, golden-buff mantle-stripes, and white underparts. In flight, broad rounded wings.

Marshes, paddies, shore cover

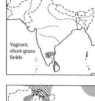

58.6. Great Snipe
Gallinago media

28cm   p.162
Rather large with buffy face and breast, white wing-bars and barred underparts with small white belly-patch. In flight, much white in outertail.

Vagrant; short-grass fields

58.2. Jack Snipe
Lymnocryptes minimus

21cm   p.163
Very small and short-billed with no pale crown-stripe; lightly marked white underparts. In flight, narrow wings and wedge-shaped all-brown tail.

Scarce; dense marsh cover

58.7. Solitary Snipe
Gallinago solitaria

30cm   p.160
Large with whiter face and broader white back-stripes than other snipes; more rufous above and warmer brown below, with yellowish legs. In flight from Pintail and Swinhoe's by pale panel on median coverts, smaller white belly-patch and more obvious rufous in much more graduated tail; flight is slow and heavy and toes do not project beyond tail.

Alpine bogs, streams

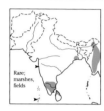

58.3. Swinhoe's Snipe
Gallinago megala

28cm   p.161
Slightly larger than Pintail; separable only in the hand. See also Common.

Rare; marshes, fields

58.8. Wood Snipe
Gallinago nemoricola

30cm   p.160
Large and rather woodcock-like; darker above than Solitary, lacking rufous, with broad golden scapular stripe. In flight, wings more rounded than other snipes, and underparts almost entirely barred.

Scarce; alpine meadows, forest-edge marsh

58.4. Common Snipe
Gallinago gallinago

26cm   p.161
Most like Pintail and Swinhoe's but in front of eye pale brow is narrower; belly and vent are white (*vs* grey). From other snipes by (usually) strong white trailing edge on secondaries and more erratic flight.

Scarce in S, E; marshes, wet fields

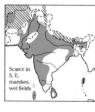

58.9. Eurasian Woodcock
Scolopax rusticola

36cm   p.160
Large; from snipes by crossbars on crown, and no bold buff stripes on upperparts. In flight, broad rounded chestnut wings. Usually silent when flushed.

Forest, edge, wet places

58.5. Pintail Snipe
Gallinago stenura

26cm   p.161
From larger Swinhoe's by little or no primary projection beyond tertials. From Common by wider buffy brow before eye, darker underwing, and indistinct pale trailing edge to wing.

Common; marshes, dry ground

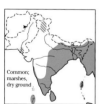

PLATE 59. SKUAS (I. Lewington)

Skuas *Catharacta* and jaegers *Stercorarius*

Gull-like seabirds with brown plumages even as adults, most species occuring in dark and pale morphs at all ages. Fly with quicker, stiffer wingbeats than gulls; harass other seabirds for food. Jaegers are smaller, slimmer, and less thick-necked than skuas, with slimmer bills. Mostly found well offshore, but sometime visit estuaries. Note pattern of white in wings, shape of tail-tip.

59.1. Parasitic Jaeger
Stercorarius parasiticus 45cm p.187

Vagrant; Pakistan coast at least. Slimmer than Pomarine with thinner, darker bill. Adult has short, pointed tail-streamers. Imm. almost uniform below, with indistinct barring on rump, and relatively dark underwing with a single pale carpal patch. See text (p. 187) for possible Long-tailed *S. longicaudus*.

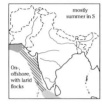
mostly summer in S

59.2. Pomarine Jaeger
Stercorarius pomarinus 56cm p.186

Heavy-bodied and deep-bellied, with heavier bicoloured bill and broader 'arm' of wing than Parasitic. Adult has diagnostic spatulate tail-streamers. Imm. typically is more barred than imm. Parasitic, with body darker than underwing, and a whitish double (*vs* single) carpal patch from below. May look more like a young gull but has conspicuous pale bars on rump.

On-, offshore, with larid flocks

Scarce; offshore, near seabird flocks

59.3. Brown Skua
Catharacta antarctica 58cm p.186

Large and bulky with a massive bill; dark brown with coarse pale streaks on mantle and flanks (more prominent when worn), and white carpal patch on both surfaces of wing (which distinguishes from dark-morph jaegers). Juv. is usually very dark warm brown (*vs* dark grey in juv. South Polar).

59.4. South Polar Skua
Catharacta maccormicki 53cm p.186

Vagrant; Sri Lanka coast at least. Smaller than Brown with shorter, lighter, less hooked bill, shorter legs, and colder, greyer overall tone. Pale morph is dark above with whitish head and underparts; dark morph and imm. are dark brown, usually with a pale hindneck and often a pale patch at base of bill and slightly barred flanks. Juv. is dark cold grey.

PLATE 60. LARGE GULLS (I. Lewington)

Gulls *Larus* (continued on Plate 61)

Gregarious waterbirds, predominantly white with black tips (often with white spots) on grey wings; juvenile plumages are usually browner and often heavily marked, especially on mantle and tail. The species on this plate belonging to the 'Herring Gull' complex (Heuglin's and Yellow-legged) present a taxonomic nightmare that has led to considerable confusion in the field and a profusion of questionable sight records, which could involve two or three additional species; see text. Many individuals of this group will defy positive identification. Note size, wing pattern, bill and leg colour.

60.1. Heuglin's Gull
Larus heuglini taimyrensis 60cm p.188

Hypothetical. Averages larger, with slightly paler mantle and slightly less black on wing-tip than nominate (thus more similar to Yellow-legged), but race not always recognised. Early regional specimens have been reassigned to nominate *heuglini*.

60.2. *Larus h. heuglini*

Bulky, with heavy bill; darker above than Yellow-legged. Soft-part colours as for Yellow-legged. In flight has large black wing-tips with small white spot on outermost primary. Non-breeder usually has heavy dark streaks on nape, unlike any Yellow-legged. First-winter has heavier dark blotches above than same-age Yellow-legged, and small dark marks on nape. Juv. has dark crown-streaks and brownish blotching below. See text for hypothetical Lesser Black-backed *L. fuscus* (p. 189).

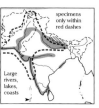

60.3. 'Steppe' Gull
Larus h. barabensis 60cm p.189

Like Heuglin's but mantle much paler and bill averages thinner. In flight, has fairly small, incurved black wing-tip with large white spot on outer two primaries. At all ages, mottling on upperparts is less than in Heuglin's, and juv. has very lightly blotched above. See text for hypothetical Mongolian *L. mongolicus* (p. 189) to which some specimens of *barabensis* were earlier assigned, and for hypothetical Armenian *L. armenicus* (p. 188).

60.4. Caspian Gull
Larus cachinnans

Hypothetical. Pale-backed, large, and long-billed compared to Heuglin's.

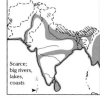

60.5. Great Black-headed Gull
Larus ichthyaetus

69cm p.189

Huge, with black subterminal band on tail, heavy angular bill and a flat crown. Breeder has black hood; non-breeder a broad dark patch from eye to hindcrown.

PLATE 61. SMALL GULLS (T. Schultz)

Gulls *Larus* (continued from Plate 60), *Rissa*

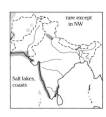
Salt lakes, coasts; rare except in NW

61.1. Slender-billed Gull
Larus genei

42cm   p.190

Very pale, with small head, pale eye, very long neck and bill; wing-pattern like Common Black-headed's. First-winter has a weak cheek-spot.

61.2. Black-legged Kittiwake
Rissa tridactyla   39cm   p.190

Vagrant; Afghanistan, Goa. Mainly oceanic. From Mew by black legs, no white spots on black wing-tip. Non-breeder has grey nape and black cheek-patch. Juv. has blackish nape-patch, tail-tip and upperwing-coverts.

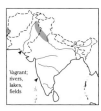
Vagrant; rivers, lakes, fields

61.3. Mew Gull
Larus canus

50cm   p.188

Smaller than Yellow-legged and Heuglin's (Pl. 60) with finer bill (no red spot) and larger, dark eye. In flight, narrow black wing-tip with large white spots on outer two primaries. Non-breeder has brown-mottled head, and may have blackish band on bill. Imm. mottled brown overall with black tail-band.

Steppe lakes; salt lakes, big rivers, coasts; common in S

61.4. Brown-headed Gull
Larus brunnicephalus

42cm   p.189

From Common Black-headed by the pale eye, heavier bill, more black on upperwing-tip. Breeder has brown hood. Non-breeder and juv. have black cheek-spot. In flight, has white triangle on outer primaries, with broad black wing-tip. See text for possible Relict *L. relictus* (p. 190).

61.5. Little Gull
Larus minutus   27cm   p.190

Vagrant; NW. Very small with fine bill and pale upperparts; dark underwing with white trailing edge. Breeder has black head and bill; non-breeder has small black ear-patch and dark smudge on crown. Juv. has black crown-patch, and dark on upper wing-coverts.

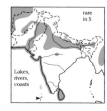
Lakes, rivers, coasts; rare in S

61.6. Common Black-headed Gull
Larus ridibundus

38cm   p.190

Smaller than Brown-headed with a smallish bill, dark eye. Large white triangle on outer primaries with short black trailing edge (no black wing-tip); from below, black patch on middle primaries. Non-breeder has blackish cheek-spot.

Desert islets; off coasts

61.7. Sooty Gull
Larus hemprichii

45cm   p.187

Long-billed, long-legged and long-winged, with dark brown upperparts and breast, dark hood and white collar. Juv. has brown breast and white underparts; paler overall than juv. White-eyed, with pale-tipped upperparts.

61.8. White-eyed Gull
Larus leucophthalmus   39cm   p.187

Hypothetical; Maldives. From larger, heavier Sooty in all plumages by long, thin, downcurved bill (may be shorter in juv.), and conspicuous broken white eye-ring.

PLATE 62. PALE *STERNA* TERNS (T. Schultz)

Terns *Sterna* (continued on Plates 63–64)

More delicately built than gulls, with narrower, more pointed wings and more buoyant, often graceful flight. Highly gregarious, mostly colonial breeders, and often highly vocal, with abrupt chattering and screeching calls. Note pattern of dark at wing-tip, soft-part colours and degree of tail-fork.

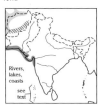

62.1. Little Tern
Sterna a. albifrons

23cm   p.195

Nominate has rather short tail streamers, white rump, dark shafts to outer primaries.

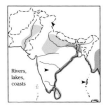

62.2. *Sterna albifrons pusilla/ sinensis*

Very small; breeder has long yellow bill (sometimes black-tipped), black cap with white forehead and short brow, and orange legs. Non-breeder and imm. have dark bill, dull legs, mainly white forecrown. Juv. is scaled above. Race *pusilla* (and especially E *sinensis*) has shining white shafts to outer primaries, and grey rump, and, in breeding plumage, longer tail-streamers.

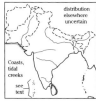

62.3. Saunders's Tern
Sterna saundersi

23cm   p.195

Breeder told from breeding Little by broader forehead-patch without brow, darker grey rump and tail, and more black on outer primaries. Inseparable from Little in other plumages.

62.4. Arctic Tern
Sterna paradisaea   36cm   p.195

Vagrant; Kashmir. From Common by shorter bill, very short legs, and tail-streamers extending beyond wing-tips. Breeder is grey below with an all-red bill; non-breeder has black bill, and may stay grey below, with reduced black on rear crown. In flight, all (*vs* just inner) primaries translucent, with sharp narrow (*vs* broad, diffuse) dark trailing edge to outer primaries from below. Juv. has greyer mantle than juv. Common or Roseate, with white forecrown, and is less strongly patterned above than juv. Roseate or Black-naped.

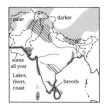

62.5. Common Tern
Sterna hirundo

36cm   p.194

At rest, has mid-length legs and bill, and tail does not project beyond wing-tips. Breeder is greyish below with dark-tipped bright red bill, and red legs. Non-breeder has black bill, dark carpal bar, and white forehead, with usually more black on rear crown than some similar species. Juv. variable; may have pale base on bill, and dark markings on upperparts, with a dark bar on secondaries, unlike juv. Arctic. In flight, note dark triangle between inner and outer primaries from above, and from below a dark trailing edge on outer primaries.

62.6. Roseate Tern
Sterna dougallii

38cm   p.194

Whiter above and below than Common and Arctic, with longer bill and tail; flies with shallower, stiffer wing-beats, no dark trailing edge to outer primaries and no grey outer web to tail feathers. Breeder often washed pink below; bill can be nearly all-black but in region is usually mostly red. Juv. like juv. Arctic and Common, but with black bill and V-shaped marks on mantle and scapulars.

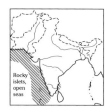

62.7. White-cheeked Tern
Sterna repressa

35cm   p.196

Like Common but dingier, with white cheek-stripe and grey rump, tail and vent. Breeder has black-tipped red bill. Non-breeder and juv. have dark bill, are darker above than equivalent Common, with broader dark carpal bar and more black on head. In flight pale mid-wing panel sits inside darker trailing edge to entire underwing.

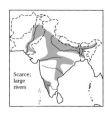

62.8. Black-bellied Tern
Sterna acuticauda

33cm   p.196

Bill always orange (tipped black in imm.), tail-streamers very long and grey. Breeder has black cap but white lores, grey breast and black belly. See Whiskered (Pl. 64). Non-breeder and imm. white below, with streaked crown and black mask. Juv. has dense blackish markings on upperparts, and lacks dark nape-patch and carpal bar of juvs. of similar species.

62.9. Black-naped Tern
Sterna sumatrana

33cm   p.194

Very pale above, with white crown and black stripe from eye to nape, long black bill and deep tail-fork. In flight, underwing white with dark leading edge to primaries.

PLATE 63. LARGE AND CRESTED TERNS (T. Schultz)

Terns *Gelochelidon, Gygis, Sterna*

63.1. Gull-billed Tern
Gelochelidon nilotica

38cm   p.192
Very pale with short, heavy black bill, black crown and legs, and relatively short forked tail. Non-breeder has white head with black around eye. Juv. similar but with pale-based bill and slight dusky scaling on buff-tinged upperparts.

63.2. White Tern
Gygis alba

29cm   p.201
Small with very weak tail-fork; white with dark eye and bill. Juv. has indistinct dusky smudges above.

63.3. River Tern
Sterna aurantia

42cm   p.194
Rather large and uncrested; dark grey above with black cap extending to lores, long tail-streamers, and heavy orange-yellow bill and reddish feet. Non-breeder has black mask and nape, and dusky bill-tip. Imm. has brown fringes above.

Crested terns *Thalasseus, Hydroprogne*

Medium to large, with shallowly forked tails, shaggy nape-crests. Note back and bill colour.

63.4. Sandwich Tern
Thalasseus sandvicensis

43cm   p.193
Pale with long, slim black bill with yellow tip, and mid-length white tail; primaries tipped black on underwing. Non-breeder has white cap with black nape-band, like Lesser Crested. Juv. rather uniformly scalloped above.

63.5. Lesser Crested Tern
Thalasseus bengalensis

43cm   p.193
Smaller and paler than Great, with orange bill; non-breeders have black restricted to hindcrown and nape. Imm. like imm. Great but wing-bars are paler.

63.6. Great Crested Tern
Thalasseus bergii

47cm   p.193
Stocky and thick-necked; darker grey above than Lesser, with stouter yellow (*vs* orange) bill that appears to droop slightly. Breeder has narrow white forehead, black crown and crest; non-breeder has blackish cheek and mottled midcrown. Imm. more extensively dark on head than imm. Lesser, with spangled upperparts, three dark bands on upperwing-coverts, and dark tail-tip. Smaller, paler-mantled race *thalassina* has been reported from Sri Lanka.

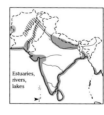

63.7. Caspian Tern
Hydroprogne caspia

51cm   p.192
Very large with huge red bill, shaggy crest, broad dark-tipped wings and short, notched tail. Non-breeder has black streaks on cap.

PLATE 64. DARK TERNS AND SKIMMER (T. Schultz)

Terns Sterna

These species are very dark above with white forehead-patches. Note back colour, amount of white in tail, extent of white in forehead.

64.1. Bridled Tern
Sterna anaethetus

37cm p.196

Dark brownish-grey above with black crown, narrow white forehead and brow. Non-breeder and imm. have white streaks on crown, pale fringes above; juv. scaled above.

Rocky islets, open seas

64.2. Sooty Tern
Sterna fuscata

43cm p.196

Black above with broad white forehead-patch, no supercilium. Juv. has dark brown head and breast, and white-scaled upperparts.

Sandy islets, open seas

Terns Chlidonias

Small with short bills, notched tails, and rather short broad wings; vent white. Buoyant flight. Note presence of nuchal collar, rump colour.

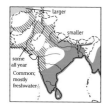

larger
smaller

64.3. Whiskered Tern
Chlidonias hybrida

25cm p.197

Breeder told from breeding White-cheeked and Black-bellied (Pl. 62) by short, notched tail and dark grey underparts with white vent. Non-breeder is grey above with black eye-stripe to nape and streaks on hindcrown, but no cheek-patch. Juv. has dark cheek-patch recalling non-breeding White-winged and Black, but note black-and-buff 'checkerboard' mantle.

some all year
Common; mostly freshwater

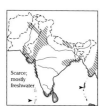

64.4. White-winged Tern
Chlidonias leucopterus

23cm p.197

Always has a white rump; note also nearly square pale grey tail (looks white). Breeder is black with largely white wings, black wing-linings, white upperwings and tail. Non-breeder has dark ear-patch narrowly connected to grey (slightly streaked) crown. Juv. has blackish-brown mantle, and darker lesser wing-coverts than non-breeder, but has diagnostic facial pattern.

Scarce; mostly freshwater

64.5. Black Tern
Chlidonias niger

23cm p.197

Hypothetical; widely sight-recorded. Grey rump and more forked tail than White-winged. Breeder is black with grey wings and tail, and white vent. Non-breeder has black crown broadly linked to cheek-patch (narrow in non-breeding White-winged), and darkish patches on breast-sides. Juv. is similar to non-breeder, but with darker mantle on lightly mottled, pale-fringed upperparts.

Noddies Anous

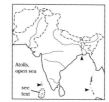

Very dark with pale caps and wedge-shaped tails.

64.6. Black Noddy
Anous minutus

37cm p.200

Blacker than congeners, with whiter crown and broad black lores; in flight, underwing blacker and tail more forked.

Atolls, open sea
see text

64.7. Lesser Noddy
Anous tenuirostris

33cm p.200

Like Black but paler and greyer, with pale grey crown. Adult has pale grey lores; juv. dark lores. In flight underwing and tail dark grey, latter little forked.

Atolls, open sea
see text

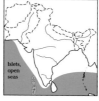

64.8. Brown Noddy
Anous stolidus

41cm p.200

Brown with pale grey crown; relatively shorter-billed and stouter than other noddies. In flight, paler wing-linings.

Islets, open seas

Skimmer Rynchopidae

Aberrant large tern-like birds with lower mandibles uniquely elongated and bill strongly laterally compressed. Sexually dimorphic in size, juvenile plumage much duller.

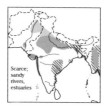

64.9. Indian Skimmer
Rynchops albicollis

40cm p.201

Long orange bill with projecting lower mandible; black with white forehead, neck and underparts. Juv. brown above with heavy pale scales and shorter black-tipped bill.

Scarce; sandy rivers, estuaries

PLATE 65. SANDGROUSE (H. Burn)

Sandgrouse Pteroclidae

Cryptic, granivorous birds of arid habitats. Flocks make long flights daily to water. Note breast, belly and underwing pattern, shape of tail.

65.1. Tibetan Sandgrouse
Syrrhaptes tibetanus

48cm　　　　　　　　　　p.201

Has tail-pin; finely barred head and breast with orange face-patch, and white belly. Female finely barred above. Juv. has less orange on face, and no tail-pin. In flight, dark underwing; above, buff on coverts with dark flight feathers.

65.2. Pallas's Sandgrouse
Syrrhaptes paradoxus

48cm　　　　　　　　　　p.202

Has tail-pin; heavily barred above with orange face and postocular streak, black belly-patch and white vent. Male grey on crown and breast with blackish breast-band; female streaked on crown with narrow black 'choker' but no breast-band. Juv. similar but breast-sides more spotted. In flight, both wing surfaces mostly pale.

65.3. Black-bellied Sandgrouse
Pterocles orientalis

39cm　　　　　　　　　　p.203

No tail-pin; black belly, 'choker' and narrow breast-band. Male has chestnut throat, and grey breast; female similar to smaller, slenderer female Chestnut-bellied. In flight, white wing-lining with black flight feathers.

65.4. Pin-tailed Sandgrouse
Pterocles alchata

38cm　　　　　　　　　　p.202

Has tail-pin; diagnostic black stripe behind eye and combination of white belly and black breast-bands (two in male, three in female). In flight, white belly and wing-lining with blackish flight feathers.

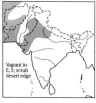

65.5. Chestnut-bellied Sandgrouse
Pterocles exustus

28cm　　　　　　　　　　p.202

Has tail-pin; black breast-band and blackish belly. Male mostly buffy; female is heavily barred above with black spots on neck, and barring on dark belly. Juv. lacks breast-band. Blackish underwing and belly diagnostic in flight.

65.6. Spotted Sandgrouse
Pterocles senegallus

36cm　　　　　　　　　　p.202

Has tail-pin; orange cheek and throat. Male is pale without black marks on head, breast or wing; female is buffy, and spotted above and on breast. Juv. paler with dense chevrons, no tail-pin. In flight, black stripe on centre of belly, with pale wing-lining, black secondaries and paler primaries.

65.7. Crowned Sandgrouse
Pterocles coronatus

28cm　　　　　　　　　　p.203

No tail-pin; pale with yellowish-orange face-patch. Male has black crescent at base of bill; female's body entirely finely barred. In flight like Spotted, but lacks dark belly-stripe.

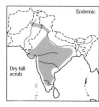

65.8. Painted Sandgrouse
Pterocles indicus

28cm　　　　　　　　　　p.203

No tail-pin. Male is heavily barred with unmarked buff face, breast and shoulder, black-and-white bars on crown, and two breast-bands. Female and juv. like equivalent Crowned but darker and more rufescent. In flight, dark with pale grey underwing.

65.9. Lichtenstein's Sandgrouse
Pterocles lichtensteinii

27cm　　　　　　　　　　p.204

No tail-pin; small and densely barred. Male has unique buff breast-patch with a double breast-band. Female greyer than female Painted, with spotted face and throat. In flight, pale body with pale underwings.

Plate sponsored by Mrs. Evelyn Bartlett.

PLATE 66. ROCK PIGEONS AND IMPERIAL-PIGEONS (I. Lewington)

Rock pigeons and Woodpigeons *Columba* (continued on Plate 67)

Open-country species (on this plate) mainly grey, mostly Himalayan except for Rock Pigeon. Note wing, rump and tail patterns.

66.1. Rock Pigeon
Columba livia

33cm p.205
Bold black wing-bars and black terminal tail-band. Dark with no white rump-patch in most of region; paler with small rump-patch in NW. Feral birds are variable; may be white, chestnut, black or patchwork.

66.4. Stock Pigeon
Columba oenas

33cm p.205
Rather dark with red-based yellow bill; no obvious wing-bar or terminal tail-band.

66.2. Hill Pigeon
Columba rupestris

33cm p.205
Paler than similar Rock, with prominent white lower back-patch and median tail-band.

66.5. Snow Pigeon
Columba leuconota

34cm p.204
White below with black head, rump and tail with striking white mid-band.

66.3. Yellow-eyed Pigeon
Columba eversmanni

30cm p.205
Smallish, with broad yellow eye-ring, brownish mantle, pale grey rump with whitish patch, and no strong tail-band; wing has three short rows of weak spots.

66.6. Common Woodpigeon
Columba palumbus

43cm p.205
Large, bulky and rather pale, with buff neck-patch; in flight, bold white wing-panel and blackish tail-band on longish tail.

Imperial-pigeons *Ducula*

Very large and simply patterned; arboreal. Note tail pattern and colour of vent.

66.7. Nicobar Imperial-pigeon
Ducula nicobarica

43cm p.217
Like Green, but mantle dark blue, vent dark brown.

66.9. Mountain Imperial-pigeon
Ducula badia

51cm p.216
Grey-brown with darker upperparts and pale terminal tail-band.

66.8. Green Imperial-pigeon
Ducula aenea

43cm p.216
Pale grey with glossy green upperparts and chestnut vent. In poor light, from Mountain by dark vent and all-dark tail; from Andaman Woodpigeon (Pl. 67) by pale belly.

66.10. Pied Imperial-pigeon
Ducula bicolor

41cm p.216
White with black wing- and tail-tip.

PLATE 67. WOODPIGEONS AND NICOBAR PIGEON (K. Williams)

Woodpigeons *Columba* (continued from Plate 66)

Large, secretive, somewhat nomadic forest pigeons. Often seen only in flight overhead, when most appear all dark. Most do not overlap in range. Note wing pattern, colour of breast *vs* belly and vent.

Forest, edge

67.1. Speckled Woodpigeon
Columba hodgsonii
38cm p.206
Mostly dark slate with a paler grey head, heavily scaled underparts, and white spots on wing-coverts. Male has maroon on mantle and belly, dark brown in female.

Endemic
Scarce; hill forest

67.4. Ceylon Woodpigeon
Columba torringtonii
36cm p.206
Dark with vinous-purple head, neck, mantle and breast, and black-and-white checkered neck-patch.

Forest

67.2. Ashy Woodpigeon
Columba pulchricollis
36cm p.206
Very dark above with pale grey head and buffy broad collar with black markings on side of neck, and dark greyish below with paler buffy belly and vent.

Scarce; woods, ravines

67.5. Pale-capped Woodpigeon
Columba punicea
36cm p.206
Dark chestnut above with striking pale crown, and slightly paler rufous below. Crown greyer in female, rufous in juv.

Endemic
Scarce; forest, plantations

67.3. Nilgiri Woodpigeon
Columba elphinstonii
42cm p.206
Grey below with paler head, and rich chestnut mantle.

Endemic
Scarce; forest
larger, head whiter

67.6. Andaman Woodpigeon
Columba palumboides
41cm p.207
Dark; pale-headed with red-based yellow bill and dark red eye-ring. From Green Imperial-pigeon (see also Pl. 66) by evenly dark grey belly and vent. Male has whitish head.

Nicobar Pigeon *Caloenas*

Strange, striking pigeon of small islands. Female slightly less elaborately plumaged than male; juvenile much less than either.

Forest, mangroves om islets

67.7. Nicobar Pigeon
Caloenas nicobarica
41cm p.212
Large; very dark glossy green with hackled upperparts and very short white tail. Juv. lacks hackles and has a dark tail. In flight, wings very broad.

PLATE 68. TURTLE-DOVES AND CUCKOO-DOVES (I. Lewington)

Turtle- and collared-doves *Streptopelia*

Generally pale doves with blackish primaries and tails, latter with broad white corners prominent in flight. Note colour of head and upperparts, details of neck-patch.

68.1. European Turtle-dove
Streptopelia turtur

28cm   p.208
Smaller and more lavender-toned below than similar western race of Oriental, with larger red eye-ring, smaller black centres to mantle feathers, and paler, browner rump.

68.2. Oriental Turtle-dove
Streptopelia orientalis (part)

33cm   p.208
Large and rufescent, with small silver-tipped black neck-patch, scaly upperparts, and grey rump. Rustier in Peninsula; browner above in NE.

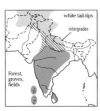

68.3. *Streptopelia orientalis meena*

NW form pale and buffy-toned, and very like European.

68.4. Laughing Dove
Streptopelia senegalensis

27cm   p.208
Small, with a longish tail and rufous-and-black checkerboard neck-patch.

68.5. Red Collared-dove
Streptopelia tranquebarica

23cm   p.209
Small and shortish-tailed with black hindcollar. Male pale clay-red with blue-grey head, rump and tail; female like small Eurasian, with white vent.

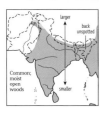

68.6. Spotted Dove
Streptopelia chinensis

30cm   p.209
Long-tailed; brown above with white-spotted black neck-patch, and pinkish-buff below. In NE, lacks pale spots on mantle.

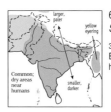

68.7. Eurasian Collared-dove
Streptopelia decaocto

32cm   p.209
Entirely pale sandy-grey with black hindcollar.

68.8. Madagascar Turtle-dove
Streptopelia picturata   p. 210

Introduced; Chagos (not illustrated). Recalls Spotted Dove but much darker, browner, and plainer, with shorter tail.

Cuckoo-doves *Macropygia*

Slim and brown with no white in long, graduated tails. Note degree of barring on underparts and tail.

68.9. Barred Cuckoo-dove
Macropygia unchall

41cm   p.211
Dark with pale face, belly and vent, and barred tail. Female barred on head, neck and breast.

68.10. Andaman Cuckoo-dove
Macropygia rufipennis

41cm   p.211
Rich bright rufous. Male is paler rufous than female, with fine bars on breast and belly; female unbarred and has stippled crown.

Zebra Dove *Geopelia*

Recalls *Streptopelia* doves but smaller, daintier, and longer-tailed. Sexes differ slightly, juvenile somewhat more.

68.11. Zebra Dove
Geopelia striata   p. 211

Introduced; Chagos (not illustrated). Small and pale, barred grey overall with longish graduated tail.

PLATE 69. EMERALD DOVE AND GREEN-PIGEONS (K. Williams)

Emerald Dove *Chalcophaps*

Small but chunky forest dove, mostly terrestrial but often seen hurtling low between trees. Sexual dimorphism weak to moderate, varying geographically; juvenile differs strongly.

69.1. Emerald Dove
Chalcophaps indica

27cm   p.211
Small and chubby, blue-green above with two pale grey bands on dark rump, and vinaceous below. Male has silver cap and shoulder. Juv. somewhat barred.

Green-pigeons *Treron*

Stocky, pale green arboreal pigeons. Females are confusing; note colour patterns of head, bill and vent, and tail shape. Nos. 2–5 are usually treated as races of one species, Pompadour Green-pigeon *Treron pompadora*.

69.2. Ashy-headed Green-pigeon
Treron phayrei

28cm   p.213
From Thick-billed by slimmer, bluish-based bill, narrow eye-ring, and brighter yellow throat. Male has orange breast-band lacking in female.

69.3. Grey-fronted Green-pigeon
Treron affinis

28cm   p.213
Throat yellow, breast green, crown grey. Female has green mantle.

69.4. Ceylon Green-pigeon
Treron pompadora

28cm   p.213
Note yellow forehead, face and throat; also dark grey-green nape and upper mantle, bronzy-olive rump and white vent with green scales. Male has dark maroon mantle, green in female.

69.5. Andaman Green-pigeon
Treron chloropterus

>28cm   p.213
Large, washed-out version of Grey-fronted with thick dark-based bill and bright yellow-olive rump. Male has limited maroon mantle with green shoulder, female's mantle green.

69.6. Thick-billed Green-pigeon
Treron curvirostra

27cm   p.214
Grey-crowned with thick red-based bill, and broad blue-green eye-ring; note whitish bars on green vent. Male has maroon mantle, some rufous on vent; female is green above.

69.7. Orange-breasted Green-pigeon
Treron bicinctus

29cm   p.212
Yellow-green crown, blue-grey nape, and mostly yellow below with rufous vent. Male has mauve and orange breast-bands; female has paler vent.

69.8. Wedge-tailed Green-pigeon
Treron sphenurus

33cm   p.216
Sturdy with longish olive wedge-shaped tail; wings look dark. Male has orange-tinged crown and breast, maroon upper mantle and shoulder and pale cinnamon vent; female has green-streaked creamy vent.

69.9. Yellow-footed Green-pigeon
Treron p. phoenicopterus

33cm   p.214
Large with grey head, belly and rump, broad green-gold collar, lilac shoulder and yellow legs.

69.10. *Treron phoenicopterus chlorigaster*

As in nominate, but with greenish-yellow belly and greyer tail.

69.11. Pin-tailed Green-pigeon
Treron apicauda

40cm/35cm   p.214
Pale greenish-yellow with long tail-pin, shorter in female. Male has orange-tinted breast.

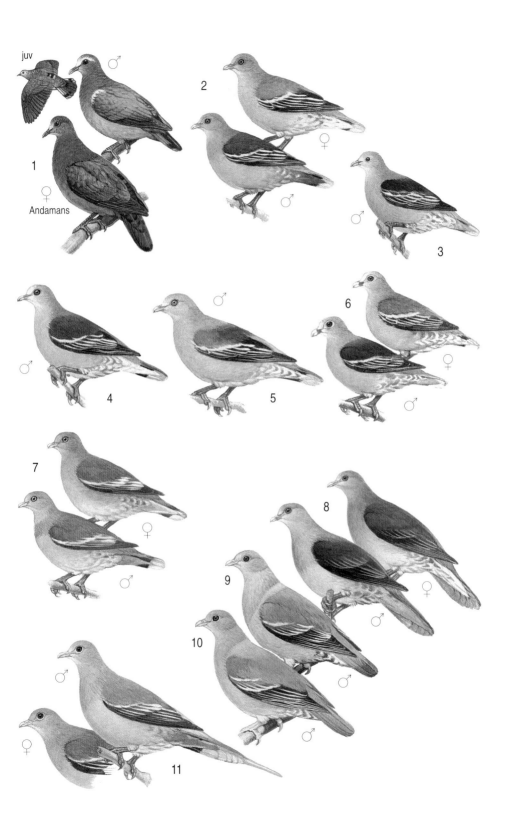

PLATE 70. HANGING-PARROTS AND SMALLER PARAKEETS (H. Burn)

Hanging-parrots *Loriculus*

Small and short-tailed, green with bright red bills and rumps; immatures greenish overall. In flight reminiscent of a small barbet.

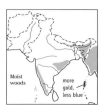

70.1. Vernal Hanging-parrot
Loriculus vernalis
14cm p.218
Bluish throat, weaker in female.

70.2. Ceylon Hanging-parrot
Loriculus beryllinus
14cm p.218
Like Vernal but with red crown, golden-tinged nape and mantle.

Parakeets *Psittacula* (continued on Plate 71)

Long-tailed and slender; species on this plate are green with differently coloured heads. Females may be very like males, or much duller like juveniles. Note wing and tail pattern, head colour.

70.3. Malabar Parakeet
Psittacula columboides
38cm p.218
Frosty bluish with blue rump, tail and primaries; dark wing has yellow-trimmed coverts. Male has red bill; female has black bill, greener belly.

70.6. Finsch's Parakeet
Psittacula finschii
36cm p.218
Smaller and yellower than Himalayan; longer tail has narrower pale tip.

70.4. Layard's Parakeet
Psittacula calthropae
31cm p.220
Short-tailed, with lavender-grey head and upperparts, broad green collar, and largely yellow-green wing.

70.7. Plum-headed Parakeet
Psittacula cyanocephala
p.219
Long blue white-tipped tail and yellowish bill. Male has magenta head and maroon shoulder-patch. Female is grey-headed with lemon-yellow neck, and no chin-strap or shoulder-patch.

70.5. Himalayan Parakeet
Psittacula himalayana
41cm p.219
Dark blue-grey head; bluer above than very similar Finsch's with darker head, and shorter, broader yellow-tipped tail.

70.8. Rosy-headed Parakeet
Psittacula roseata
36cm p.220
From Plum-headed by yellowish tip to shorter, more turquoise tail; maroon shoulder in both sexes. Male has pink head; female has paler head with ochre collar.

70.9. 'Intermediate Parakeet'
Psittacula 'intermedia' p.219, 605
Hybrid Plum-headed × Himalayan, once thought a valid species; see text.

PLATE 71. RINGED AND LARGER PARAKEETS (H. Burn)

Parakeets *Psittacula* (continued from Plate 70)

Long-tailed and slender; most are green with black chin-strap and reddish bills. Females similar to or slightly duller than males; juveniles usually duller still.

71.1. Rose-ringed Parakeet
Psittacula krameri

42cm   p.218
Bright yellow-green with no red on shoulder. Male has narrow black-and-pink collar.

71.4. Nicobar Parakeet
Psittacula caniceps

61cm   p.221
Large and bright yellow-green with pale grey head, heavy black lores and chin-strap; very dark primaries but no shoulder-patch. Female has black bill.

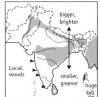

71.2. Alexandrine Parakeet
Psittacula eupatria

53cm   p.218
Large version of Rose-ringed, with large bill and maroon shoulder-patch.

71.5. Derby's Parakeet
Psittacula derbiana

46cm   p.220
Hypothetical; N Arunachal. Large version of Red-breasted, with pale lilac head and breast.

71.3. Red-breasted Parakeet
Psittacula alexandri

38cm   p.221
Grey head, heavy black chin-strap and lores, salmon-pink breast, and golden wing-patch; female has black bill. Juv. has greenish head and breast.

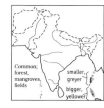

71.6. Long-tailed Parakeet
Psittacula longicauda

47cm   p.221
Rather large, with diagnostic reddish face between green crown and breast; blue in primaries and tail. Male has very long tail; female is duller with pinkish cheek and blackish bill.

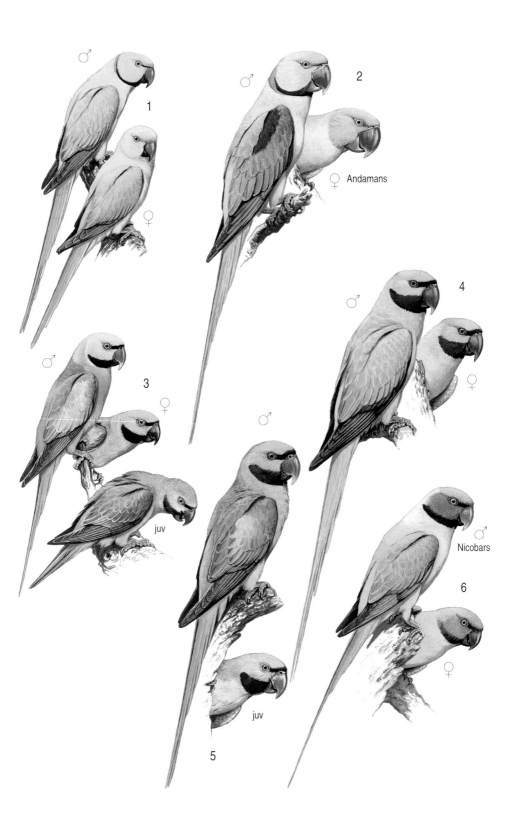

PLATE 72. SMALL CUCKOOS, CRESTED CUCKOOS AND KOEL (J. Anderton)

Glossy cuckoos *Chrysococcyx*

Small cuckoos with barred bellies and in flight broad white wing-bands. Males have vivid glossy upperparts, reduced and replaced mostly with rufous in females and juveniles. Note colour of upperparts and bill; in females, facial pattern and weight of barring on face and underparts.

72.1. Violet Cuckoo
Chrysococcyx xanthorhynchus
17cm p.227
Orange bill with no dark tip. Male is purple above; female is told from female Asian Emerald by white brow and face. Juv. is brighter rufous than female and barred above with banded tail; from Banded Bay by shorter tail and lack of dark eyeline.

72.2. Asian Emerald Cuckoo
Chrysococcyx maculatus
18cm p.226
Dark-tipped yellow bill and unbarred green central tail feathers at all ages. Male is glossy emerald above; female and juv. are similar to female Violet, but have rufous face and crown, and closer barring on underparts.

Plaintive cuckoos *Cacomantis*

Small, slim and dark-eyed; tend to perch more upright than the similar but larger *Cuculus* species (Pl. 73). Note belly colour; facial pattern in immatures.

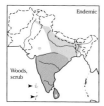

72.3. Grey-bellied Cuckoo
Cacomantis passerinus
23cm p.228
Dark grey with whitish vent. Hepatic female has unmarked tail.

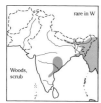

72.4. Plaintive Cuckoo
Cacomantis merulinus
23cm p.227
Paler than Grey-bellied with rufous belly; hepatic female is heavily barred on rump and tail, with rufescent face and throat. Juv. has streaked head.

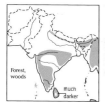

72.5. Banded Bay Cuckoo
Cacomantis sonneratii
24cm p.227
Longish bill with dark eyeline and whitish brow; rufous above, white below and entirely barred black.

Drongo-cuckoos *Surniculus*

Rather small black cuckoos, easily passed over for drongos. At least two species in region but distributions unclear. Sexes similar, juvenile differs strongly.

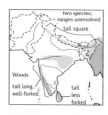

72.6. 'Fork-tailed' Drongo-cuckoo
Surniculus [*lugubris*] *dicruroides*
27cm p.228
Black with splayed, forked tail; from drongos (Pl. 175) by lazy posture, fine bill, and white bars on underside of tail. Juv. heavily speckled with white.

72.7. 'Square-tailed' Drongo-cuckoo
Surniculus lugubris   25cm p.228
Map above. From Fork-tailed by notched, less flared tail-tip. Taxonomy and ranges of drongo-cuckoos not yet clear. Juv. as for Fork-tailed.

Crested cuckoos *Clamator*

Showy, with black upperparts, long black crests and tails. Note wing colour and pattern, presence of tail-tips.

72.8. Jacobin Cuckoo
Clamator jacobinus
31cm p.225
Small, with white carpal patch and tail-tips. Juv. brown above, fulvous below.

72.9. Chestnut-winged Cuckoo
Clamator coromandus
42cm p.225
White half-collar, chestnut wings, and rufous throat. Juv. scaled rufous above.

Koel *Eudynamys*

Large, long-tailed, noisy cuckoo that parasitises crows. Sexes differ greatly; juveniles more like female.

72.10. Asian Koel
Eudynamys scolopaceus
43cm p.226
Red eyes, hooked greenish bill, and heavy rounded tail. Male is black; female heavily spotted white above with barred underparts and tail. Larger in NE and Andamans, where female rufescent.

PLATE 73. TYPICAL CUCKOOS AND HAWK-CUCKOOS (J. Anderton)

Cuckoos *Cuculus*

Slimmer and longer-winged than hawk-cuckoos; generally unmarked grey above and barred below; most have hepatic female form. Juveniles have whitish nape-patches and dark eyes. Note diagnostic songs; eye colour, tail/rump contrast, boldness of barring below and tail pattern.

Hawk-cuckoos *Hierococcyx*

Recall accipiters (Pl. 22) in pattern as adults and juveniles. From *Cuculus* by shorter wings, broadly barred tails, and stronger throat/cheek contrast. Juveniles are brown and heavily streaked; immatures intermediate. Note degree of streaking on breast, barring on belly, colour of mantle and tail-tip, tail pattern.

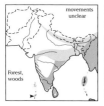

73.1. **Indian Cuckoo**
Cuculus micropterus

33cm   p.230

Brownish above with broad blackish tail-band, dark eye, and broad dark bars below; no hepatic plumage. Juv. is brownish and blotchy, with irregular pale patches on head.

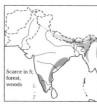

73.5. **Large Hawk-cuckoo**
Hierococcyx sparverioides

38cm   p.229

Larger and darker than Common with black chin, browner upperparts, chestnut breast with dark streaks and heavier barring on flank; subadult dark brown above, coarsely streaked below.

73.2. **Common Cuckoo**
Cuculus canorus

33cm   p.231

Paler above than Oriental with finer bars below and whiter vent. Hepatic female has rather narrow, widely spaced bars above, with scarcely marked rump.

73.6. **Common Hawk-cuckoo**
Hierococcyx varius

34cm   p.229

Smaller, paler than Large with softer rufous breast and bars on belly, and clearer bars on tail. Subadult is paler and less boldly marked than imm. Large. Sri Lanka race *ciceliae* is dark and heavily marked, thus similar to allopatric Large.

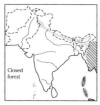

73.3. **Oriental Cuckoo**
Cuculus saturatus

31cm   p.230

Darker above and on tail than Common, with paler rump, and buffier below with broader bars; plain *vs* barred inner primary coverts on underwing. Hepatic female very heavily barred above (including on rump). Normal juv. dark above with narrow white scales, and whitish below.

73.7. **Whistling Hawk-cuckoo**
Hierococcyx nisicolor

29cm   p.229

Dark blue-grey above, with orange eye, blackish chin, unbarred white belly and rufous tail-tips. Juv. blackish above with narrow whitish fringes; large black droplet-like spots below.

73.4. **Small Cuckoo**
Cuculus poliocephalus

26cm   p.230

Small with a dark eye and rump. Hepatic female (not illustrated) is very dark rufous and lightly barred above.

PLATE 74. MALKOHAS AND COUCALS (J. Anderton)

Malkohas *Phaenicophaeus, Taccocua*

Long-tailed, heavy-billed and round-winged arboreal cuckoos; all have large white tail-tips, all but Sirkeer have pale green bills. Mostly silent. Note colour of underparts.

Coucals *Centropus*

Large and awkward, rather terrestrial cuckoos with stout bills, red eyes, heavy tails and rounded rufous wings. Note bill length and colour, mantle colour.

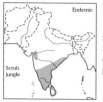

74.1. Blue-faced Malkoha
Phaenicophaeus viridirostris

39cm　　　　　　　　p.224
Smaller than Green-billed, with blue eye-patch, dark throat and breast, and pale belly and vent.

74.2. Green-billed Malkoha
Phaenicophaeus tristis

51cm　　　　　　　　p.225
Extremely long-tailed with red eye-patch, grey throat and breast, and blackish vent.

74.3. Red-faced Malkoha
Phaenicophaeus pyrrhocephalus

46cm　　　　　　　　p.224
Black above with heavy pale bill, bare red face, and white belly; tail looks nearly all white from below.

74.4. Sirkeer Malkoha
Taccocua leschenaultii

43cm　　　　　　　　p.224
Pale brown with bushy crest, curved red bill, blackish eyeline and orange-buff belly.

74.5. Lesser Coucal
Centropus bengalensis

33cm　　　　　　　　p.223
Relatively small, short-tailed and stubby-billed; wing-lining chestnut vs black, and eye dark. In breeding, black with dull rufous mantle and wings, pale shaft-streaks in wing-coverts, and bronzy-green tail. Non-breeding and imm. are brown above with pale shaft-streaks, and buffy below; rufous wings barred black in subadult. Juv. is shorter-tailed, very rufous and spotted black above with weak shaft-streaks.

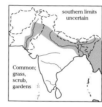

74.6. Greater Coucal
Centropus sinensis (part)

48cm　　　　　　　　p.222
Very large with entirely chestnut mantle and wings; black plumage glossed blue-purple. Juv. has pale bars overall.

74.7. 'Southern' Coucal
Centropus [sinensis] parroti

From very similar Greater by blue-green gloss to black plumage, dull brownish forehead and throat, and slightly less extensive rufous on mantle. Juv. is duller and unbarred, with darker mantle and wings; very similar to adult and juv. Green-billed.

74.8. Andaman Coucal
Centropus [sinensis] andamanensis

48cm　　　　　　　　p.223
Large and brown, with dark chestnut wings; head is variably paler.

74.9. Green-billed Coucal
Centropus chlororhynchos

43cm　　　　　　　　p.222
From 'Southern' by pale green bill, and deeper chestnut mantle not contrasting with purplish-black head. Imm. has duller soft-part colours; very like juv. 'Southern.'

PLATE 75. HAWK-OWLS AND BARN-OWLS (L. McQueen)

Hawk-owls *Ninox*

Smallish, small-headed and earless hawk-like owls with reduced facial disks, yellow eyes and longish, banded tails.

75.1. Brown Hawk-owl
Ninox scutulata

32cm   p.247

Mid-sized, with small white spot over bill, white scapulars (often hidden), rufous streaks on breast grading to chevrons on whitish belly, and whitish tail-tip. Pale in NW, and much darker in NE and S. Rufescent below in S Nicobars; C Nicobars with rufescent streaking.

75.2. Andaman Hawk-owl
Ninox affinis

28cm   p.248

Small and much paler than sympatric Hume's, with unmarked vent; similar to Brown, but with much white on face and bold chestnut streaks (*vs* chevrons) on belly.

75.3. Hume's Hawk-owl
Ninox obscura

32cm   p.248

Large; very dark brown and almost unmarked: small white spot over bill, broad white bars on vent, and broad blackish bands on tail with narrow whitish tip.

Bay owls *Phodilus*

Rather small, very short-tailed and short-legged owls with rather heart-shaped facial disk and dark eyes, dark chestnut upperparts and vinous-tinged underparts. Sexes and ages similar. Songs elaborate, eerie whistles very unlike other owls. Formerly considered a single species.

75.4. Oriental Bay Owl
Phodilus badius

29cm   p.234

Bright chestnut above and sparsely speckled, with weakly banded wings and tail.

75.5. Ceylon Bay Owl
Phodilus assimilis

29cm   p.235

Dark chestnut and golden above and densely speckled; heavily banded wings and tail.

Barn and grass-owls *Tyto*

Medium-sized, very long-legged owls with rather heart-shaped facial disk and dark eyes, finely speckled, brownish upperparts and paler speckled underparts. Females and juveniles usually darker, especially on facial disk, than males. Lack obvious song.

75.6. Andaman Barn-owl
Tyto deroepstorffi

34cm   p.233

Chunky; dark brown above with vinous facial disk, rufous speckles, and short barred tail, and golden-rufous speckled black below.

75.8. Eastern Grass-owl
Tyto longimembris

36cm   p.233

Darker above than Common Barn-owl with dark bands on wings and tail. Facial disk whitish in male, buffy in female, vinous in juv. In flight wings pale with dark carpal patch and tips.

75.7. Common Barn-owl
Tyto alba

36cm   p.233

Very pale with white facial disk and underparts, grey-and-gold upperparts and unbanded tail. Regionally rare dark morph has vinous facial disk and buffy underparts.

PLATE 76. FISH-OWLS, SNOWY OWL AND EAGLE-OWLS (L. McQueen)

Very large owls, all closely related and sometimes placed in a single genus *Bubo*. Eartufts prominent in most, and songs very low-pitched, often duetted.

Fish-owls *Ketupa*

Partly diurnal; similar to eagle-owls, but plainer, with recumbent ear-tufts, weakly defined facial disks, and partly bare legs.

76.1. **Tawny Fish-owl**
Ketupa flavipes

56cm   p. 241

Large, with black bill, unmarked tawny facial disk, boldly barred wings and tail, and boldly streaked underparts. In flight, wings and tail boldly banded on both surfaces.

76.2. **Brown Fish-owl**
Ketupa zeylonensis

61cm   p. 240

Large, grey-brown and finely streaked overall with white spots on scapulars. In flight, wings and tail moderately banded on both surfaces.

76.3. **Buffy Fish-owl**
Ketupa ketupu

50cm   p. 242

Much smaller and duller than Tawny, less boldly patterned above and below, with unstreaked warm brown ear-coverts; and a small white crescent above bill.

Eagle-owls *Bubo*

Large with long, widely separated upright ear-tufts and well-defined facial disk.

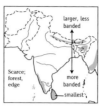

76.4. **Forest Eagle-owl**
Bubo nipalensis

63cm   p. 240

Very large and pale, with dark eyes in whitish facial disk, broad white brow running onto long barred ear-tufts, and black chevrons on whitish underparts. Smaller and darker, more barred in Sri Lanka. Juv. whitish and narrowly barred overall.

76.5. **Dusky Eagle-owl**
Bubo coromandus

58cm   p. 240

Uniform grey-brown, with yellow eyes, narrow border to plain grey facial disk, narrow black streaks below, and broad whitish scapular edges. In flight, wings and tail boldly banded below, with no obvious dark carpal patch.

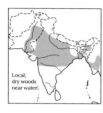

76.6. **Indian Eagle-owl**
Bubo bengalensis

54cm   p. 239

Heavily marked, with black border to plain pale facial disk, largely blackish mantle and tertials, pale scapular spots, and bold tail bands. Dark carpal in flight.

76.7. **Eurasian Eagle-owl**
Bubo bubo

56cm   p.239

Huge; larger and paler than Indian, with weaker border to facial disk, mainly tawny mantle, mainly pale tertials, and weakly banded wings and tail.

Snowy Owl *Nyctea*

A huge white owl closely related to eagle-owls but lacking ear-tufts. Irruptive in hard winters.

76.8. **Snowy Owl**
Nyctea scandiaca   61cm   p. 242

Vagrant; NW Pakistan. Large, white and earless; female and imm. are scaled with black except on face and breast.

PLATE 77. *ASIO* AND WOOD-OWLS (L. McQueen)

Owls *Asio*

Slim, medium-sized owls with distinct facial disks and long wings.

Scarce, erratic; conifers

77.1. Northern Long-eared Owl
Asio otus
37cm   p.249
Slim, with tall close-set ear-tufts, bright buff facial disk, and heavy streaks below.

rare in S
Erratic; grassland, semi-desert

77.2. Short-eared Owl
Asio flammeus
38cm   p.249
Buffy with heavy dark streaks below; yellow eyes set in black face-patches. In flight shows black carpal patch on pale underwing. Often diurnal.

Wood-owls *Strix*

Fairly large earless owls, most with dark eyes, rounded wings and rather long tails.

77.3. Hume's Owl
Strix butleri   36cm   p.244
Hypothetical; SW Pakistan. Very pale tawny and almost unmarked, with bright orange eyes.

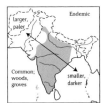
larger, paler   Endemic
Common; woods, groves   smaller, darker

77.4. Mottled Wood-owl
Strix ocellata
48cm   p.242
Pale and densely dark barred overall, including facial disk, with dark eyes and white throat-patch; rufous splashes on cheeks, crown and underparts. In flight, bands weak.

browner   more rufous
Montane forest

77.5. Himalayan Wood-owl
Strix nivicola
46cm   p.244
Mid-sized, dark and dark-eyed with pale X on face, mottled and unstreaked above with double pale wing-bars and broadly banded tertials, and coarsely streaked and cross-barred below. Paler and browner in W Himalayas; very dark or rufous in C and E Himalayas. In flight, chunky with rounded, banded wings.

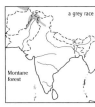
a grey race
Montane forest

77.6. Tawny Owl
Strix aluco
46cm   p.243
Like Himalayan Wood-owl but paler and greyer, and streaked (*vs* mottled) above with almost unbarred tail.

large, no rufous, barred face
Forest   smaller, face rufous
smallest

77.7. Brown Wood-owl
Strix leptogrammica
50cm   p.242
Solid brown on crown and mantle, and densely barred below with conspicuous facial disk and dark eyes. Very large in Himalayas, with barred facial disk; smaller in Peninsula with unbarred ochraceous facial disk. Juv. has white head for a few months. See text for hypothetical Spotted *S. seloputo* (p.243).

PLATE 78. SCOPS-OWLS (L. McQueen)

Scops-owls *Otus*

Small, cryptic, nocturnal owls with ear-tufts and prominent facial disks; some species have morphs or are highly variable.

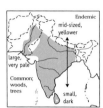

78.1. Indian Scops-owl
Otus bakkamoena

24cm   p.237
Variable; usually with dark eyes and bill-tip, black-bordered facial disk, bicoloured ear-tufts, pale upperparts with strong blackish streaks, buff scapulars and blackish crown. Very like Collared. In NW Himalayas large, heavily marked grey; in NW deserts, pale and streaked, with orange eyes. Small in SW Peninsula/Sri Lanka and often ochraceous, eyes sometimes yellowish.

78.2. Collared Scops-owl
Otus lettia

23cm   p.239
Almost indistinguishable from Indian but has pale bill-tip, concolorous crown and mantle, spotted ear-tufts. Form depicted exceptionally distinctive.

78.3. Pallid Scops-owl
Otus brucei

21cm   p.236
Pale grey with long dark streaks (heavier on crown than mantle), no rufous in plumage, and buffy scapular spots. First-winter is paler and sandier with weaker marks.

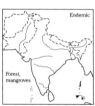

78.4. Nicobar Scops-owl
Otus alius

c.21cm   p.237
Rather large, warm brown and finely barred overall, with broad streaks on crown and very little streaking below; eyes pale yellow. See also back cover.

78.5. Andaman Scops-owl
Otus balli

19cm   p.235
Much like Mountain with larger bill and partly bare tarsi; ground colour more uniform, but more black speckling below. Eyes probably yellow.

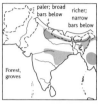

78.6. Oriental Scops-owl
Otus sunia

19cm   p.236
Small with yellow eyes; weakly streaked above but heavily below and on crown. Rufous morph more lightly streaked; brown and grey morphs have rufous in ear-tufts and face-border, and barred tertials.

78.7. *Otus sunia* S races

Sri Lanka form very small and dark; somewhat more finely marked overall in Andamans. Nicobars form mostly or entirely rufous.

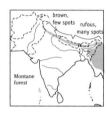

78.8. Mountain Scops-owl
Otus spilocephalus

19cm   p.236
Small with yellow eyes; unstreaked plumage with small white spots, weakly defined facial disk, white-frosted forehead, and whitish central belly-stripe. Grey-brown in W, more rufous in E.

78.9. Eurasian Scops-owl
Otus scops

19cm   p.236
Mainly grey with rufous accents over eyes and on scapulars, pale grey band on sides and rear of crown, spotted and streaked greyish ear-tufts.

78.10. Serendib Scops-owl
Otus thilohoffmanni

17cm   p.235
Very small with orange-yellow eyes, no ear-tufts, and short tail; rufous lightly speckled with black. See also back cover.

PLATE 79. OWLETS AND BOREAL OWL (L. McQueen)

Owlets *Athene, Heteroglaux*

Small and chunky earless brown owls, heavily spotted and/or streaked, with yellow eyes. Highly diurnal. Note pattern of face and underparts, spotting on upperparts.

79.1. Little Owl
Athene noctua

23cm   p. 245

Bulkier than Spotted Owlet with streaked forehead and underparts, and broad whitish tail-bands. Larger and darker in N Himalayas, paler in NW. Juv. more subtly marked.

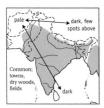

79.2. Spotted Owlet
Athene brama

21cm   p. 246

Small, heavily spotted above and scaled below, with dark cheeks enclosed by white border; tail narrowly banded. In NW can be very pale, in S and NE very dark. Juv. has little or no spotting on crown.

79.3. Forest Owlet
Heteroglaux blewitti

23cm   p. 247

Stouter and darker than Spotted, with unspotted crown and mantle, dark brown breast and broadly barred flanks on white underparts, and more boldly barred tail. Juv. has fuzzy, vague brown streaks on white underparts.

Boreal Owl *Aegolius*

Chunkier and shorter-legged than *Athene* owlets, with perpetually surprised look.

79.4. Boreal Owl
Aegolius funereus

25cm   p. 247

Small, brown above with white spots, brown mottling below, with whitish face bordered broadly with black. Juv. mostly solid dark brown.

Owlets *Glaucidium*

Small, earless, largely diurnal owls with reduced facial disks; some species have colour morphs, and juvenile plumages distinctive.

79.5. Jungle Owlet
Glaucidium radiatum (part)

20cm   p. 245

Whole plumage closely barred.

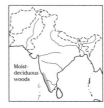

79.6. *Glaucidium radiatum malabaricum*

Like nominate but much more rufescent, especially on head.

79.7. Chestnut-backed Owlet
Glaucidium castanonotum

19cm   p.245

Darker than Jungle, with chestnut on mantle and wings, and streaked flanks.

79.8. Asian Barred Owlet
Glaucidium cuculoides

23cm   p.244

Relatively large with heavily streaked flanks and belly, and broadly banded tail. More rufous in NE; juv. has more spotted crown, and plainer mantle.

79.9. Collared Owlet
Glaucidium brodiei

17cm   p.244

Tiny version of Asian Barred, with bold 'owl face' on nape, narrowly banded tail, and crown more spotted than barred. Grey morph has narrow pale bands, broader in rufous morph. Juv. is more uniform above with pale collar, blurrier pattern below.

PLATE 80. FROGMOUTHS AND NIGHTJARS (J. Anderton)

Frogmouths *Batrachostomus*

Nightbirds with broad bills and rounded wings lacking pale patches. Perch upright, and forage within foliage.

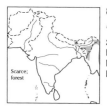

80.1. Hodgson's Frogmouth
Batrachostomus hodgsoni
27cm   p.250
Large with bold white scapular spots. Male is coarsely barred; female has large white spots on breast.

Endemic
80.2. Ceylon Frogmouth
Batrachostomus moniliger
23cm   p.249
Smaller, more uniform than Hodgson's. Male brown to grey, with pale, vermiculated scapulars; female chestnut with small white spots below.

Nightjars *Caprimulgus, Eurostopodus*

See Plate 81 for flight illustrations and additional maps

Long-winged aerial-feeding nightbirds with tiny bills. Cryptically patterned, usually with white panels in wing and tail of males, buffy and reduced in females and juveniles. Note overall colour, weight of streaks on upperparts and barring on tail.

Desert

80.3. Egyptian Nightjar
Caprimulgus aegyptius
25cm   p.253
Pale and relatively unmarked above; longer-tailed than Sykes's, buffier than European.

80.8. Grey Nightjar
Caprimulgus jotaka   29cm   p.252
Darker and richer than Indian Jungle, with broader black bars on tail and rufous spots on darker wing. Markings above very heavy, obscuring pattern.

80.9. Indian Little Nightjar
Caprimulgus asiaticus   24cm   p.254
Small and short-tailed with golden hindcollar and pale throat-spots (not band).

Scarce; semi-desert, dry woods

80.4. Sykes's Nightjar
Caprimulgus mahrattensis
23cm   p.253
Small, pale and weakly marked; shorter-tailed than Egyptian.

80.10. Jerdon's Nightjar
Caprimulgus atripennis   26cm   p.254
Shorter-tailed than Large-tailed with brown hindcollar, smaller white throat-patch and paler breast.

80.11. Large-tailed Nightjar
Caprimulgus macrurus   30cm   p.253
Large, long-tailed with black-and-gold scapular stripes, yellowish hindcollar, large unbroken white throat-patch above barred breast, and pale bars on wing-coverts.

a pale race

Rocky hills

80.5. European Nightjar
Caprimulgus europaeus
25cm   p.252
Quite pale with buff scapular stripes, whitish submoustachial and large white neck-spot. More heavily marked above than Egyptian and Sykes's but less than Indian Jungle.

80.12. Savanna Nightjar
Caprimulgus affinis   25cm   p.255
Dusky and without strong streaks or spots above; note buffy scapular stripe and pale throat-patches.

80.13. Andaman Nightjar
Caprimulgus andamanicus   25cm   p.254
Very dark, mid-sized nightjar without prominent hindcollar or scapular markings.

Almost unknown; sandy scrub jungle

80.6 Vaurie's Nightjar
Caprimulgus centralasicus
Possible; extreme NW (not illustrated). Small and very pale; warmer-toned than Sykes's and Egyptian, with streaked crown.

80.14. Great Eared-nightjar
Eurostopodus macrotis   39cm   p.251
Very large and dark, with long barred wings and tail, pale crown, unbroken white throat-band and blackish breast.

80.7. Indian Jungle Nightjar
Caprimulgus indicus   29cm   p.251
Grey above with narrow black bars on tail, black streaks on pale scapulars, and white throat-spots. Much finer-marked above than Grey, coarser than European.

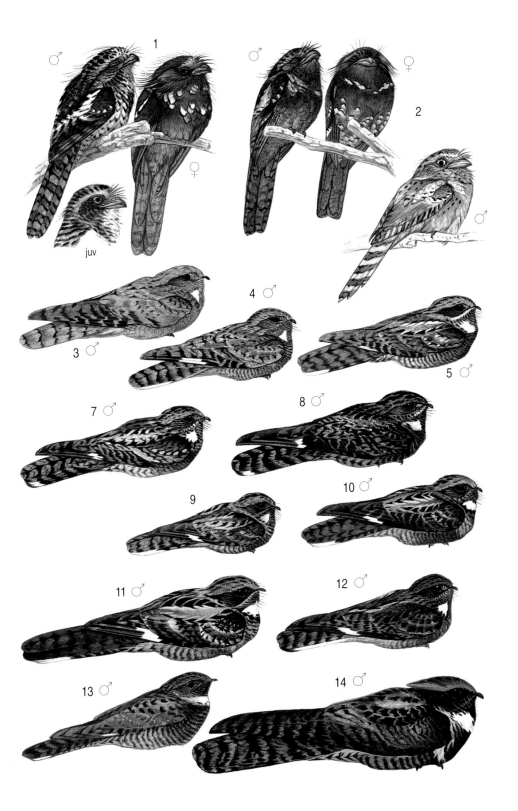

PLATE 81. NIGHTJARS IN FLIGHT (J. Anderton)

Nightjars *Caprimulgus, Eurostopodus*

See Plate 80 for non-flight illustrations and additional maps

81.1. Egyptian Nightjar
Caprimulgus aegyptius p.253

Palest regional nightjar. Underside of wing and tail pale, lightly barred and without white patches.

81.2. Sykes's Nightjar
Caprimulgus mahrattensis p.253

Rather oval patch on wing, and broad pale tail-corners. Marks white in male, buffy in female.

81.3. European Nightjar
Caprimulgus europaeus p.252

Mid-sized. Male has narrow white wing-patches and large white corners on tail; much reduced or lacking in female (not shown). Similar Indian Jungle and Grey males have white subterminal tail-band.

81.4. Indian Jungle Nightjar
Caprimulgus indicus p.251

Paler than Grey. Male has white subterminal tail-band across all but central rectrices, and narrow white wing-spots; female has reduced buff wing-spots and no tail-band.

81.5. Grey Nightjar
Caprimulgus jotaka p.252

Very dark; otherwise similar to Indian Jungle, although female wing-patches more reduced.

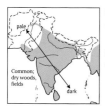

81.6. Indian Little Nightjar
Caprimulgus asiaticus p.254

Small and short-tailed; pattern like Jerdon's.

81.7. Jerdon's Nightjar
Caprimulgus atripennis p.254

Like Large-tailed but shorter-tailed; sex-for-sex pattern similar but with smaller pale patches in wing and tail.

81.8. Large-tailed Nightjar
Caprimulgus macrurus p.253

Large and long-tailed, with large white wing-spots and tail-corners in male, buffy in female.

81.9. Savanna Nightjar
Caprimulgus affinis p.255

Male has narrow white wing-spot and diagnostic all-white outertail feathers; female has buff wing-spots and no pale in tail.

81.10. Andaman Nightjar
Caprimulgus andamanicus p.254

Like Large-tailed but shorter-tailed; sex-for-sex pattern similar but with smaller pale patches in wing and tail.

81.11. Great Eared-nightjar
Eurostopodus macrotis p.251

Very large; no white in wings or tail.

PLATE 82. TREESWIFT AND SWIFTS I (J. Schmitt)

Treeswift Hemiprocnidae

Distinctive perching swift with loud, parakeet-like calls. Slightly dimorphic in plumage, juvenile very different from adult.

82.1. Crested Treeswift
Hemiprocne coronata

23cm Pl. 100, p.261

Only swift to perch upright, when diagnostic frontal crest visible. Wings and forked tail exceedingly long. In flight, larger than similarly shaped Asian Palm, and greyer with a whitish belly and white face-stripes. Male has rufous cheeks, dark grey in female. Juv. heavily scaled above and below; wings and tail shorter. Perched figures not to scale.

Swifts Apodidae

Compact, usually short-tailed birds normally seen only in flight, with rapid wing-beats (interspersed with glides) on stiff wings that immediately distinguish them from the generally more colourful swallows. Many species are very similar and extremely difficult to identify; juveniles are similar to adults.

Palm-swift *Cypsiurus*

Very small, extremely attenuated swift; in groups around palms.

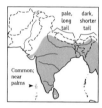

82.2. Asian Palm-swift
Cypsiurus balasiensis

13cm p.258

Tiny with very long narrow wings and tail-streamers (usually tightly folded and needle-like); browner than much larger Crested Treeswift but similarly shaped in flight. Entirely grey-brown with diffusely paler throat, more conspicuous in much darker and shorter-tailed NE form.

Swiftlets *Aerodramus, Collocalia*
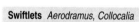

A group of very similar species, all with rather plain dark plumage and slightly notched tails; some identifiable only in the hand. Note contrast of rump, colour of belly, tail shape.

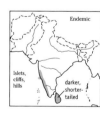

82.3. Indian Swiftlet
Aerodramus unicolor

12cm p.256

Dark with blackish wing-lining; very like larger Himalayan (with no range overlap) but rump little if any paler.

82.4. Edible-nest Swiftlet
Aerodramus fuciphagus

12cm p.257

Small and very dark, with less deeply forked tail than Himalayan.

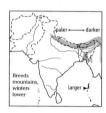

82.5. Himalayan Swiftlet
Aerodramus brevirostris

14cm p.257

Rather large with fairly deep tail-fork; dark greyish with a diffuse paler rump-band, pale scales on vent, and blackish wing-linings.

82.6. White-bellied Swiftlet
Collocalia esculenta

10cm p.256

Tiny with very slightly forked broad tail; glossy greenish-black above including rump, with white belly.

82.7. Black-nest Swiftlet
Aerodramus maximus 14cm p. 256

Rejected from avifauna (Appendix 2); regional records based on misidentified specimens.

Spinetail *Zoonavena*

A small, short-tailed swift with distinctive markings, but see Little.

82.8. Indian White-rumped Spinetail
Zoonavena sylvatica

11cm p.257

Small and broad-winged; blue-black above with bold white rump, and grey below grading to white belly and vent.

PLATE 83. SWIFTS II (J. Schmitt)

Swifts *Apus, Tachymarptis*

Mid-sized to large swifts with longer, more deeply forked tails than swiftlets or spinetails (except in Little). Note colour of rump, belly, vent and throat, shape of tail.

83.1. Little Swift
Apus affinis

15cm p.260
Small and compact, with bold white throat and rump; and shortish, squared to slightly forked tail. Darker in E Himalayas and NE with smaller and better-defined white throat-patch, narrower white rump-band, and more clearly notched tail. Compare Indian White-rumped Spinetail (Pl. 82.8).

83.2. Pallid Swift
Apus pallidus

17cm p.259
Very like Common, but overall paler and browner; see text.

83.3. Dark-rumped Swift
Apus acuticauda

17cm p.260
Like Common but tail more deeply forked and throat pale grey; scaled white below, usually more than in Pacific, with black vent.

83.4. Common Swift
Apus apus

17cm p.259
Very dark with with inconspicuous whitish throat and dark rump.

83.5. Pacific Swift
Apus pacificus

18cm p.260
Slim and blackish with bold white rump, and very deeply forked tail; usually has inconspicuous white scales on underparts, including vent.

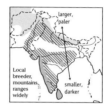

83.6. Alpine Swift
Tachymarptis melba

22cm p.259
Large and long-winged; white below with diagnostic brown breast-band; vent brown.

Needletails *Hirundapus*

Very large swifts with (usually invisible) projecting spines on short tails, a pale patch in centre of back, and prominent white vent extending up flanks in V. Note throat colour, and tail shape.

83.7. White-throated Needletail
Hirundapus caudacutus

20cm p.258
From congeners by snowy-white throat-patch. Often has white spots on tertials.

83.9. Brown-throated Needletail
Hirundapus giganteus

23cm p.258
Dark throat, with white lore-spot; tail rather longer and more wedge-shaped than in congeners.

83.8. Silver-backed Needletail
Hirundapus cochinchinensis

20cm p.258
Similar to White-throated, but pale grey throat lacks contrast with dark underparts; note very pale back-patch.

PLATE 84. HOOPOE, ROLLERS AND TROGONS (J. Anderton)

Rollers *Coracias, Eurystomus*

Stocky, heavy-billed birds with broad, boldly patterned wings and squared tails; powerful fliers of open habitats in which they make aerial forays from exposed perches. Often recall raptors or crows when perched. Juveniles duller but recognisable. Note colour of breast, pattern of wings and tail.

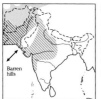

84.1. European Roller
Coracias garrulus

31cm p.270
Entirely turquoise head and underparts, with tawny mantle; in flight note dark unbanded flight feathers.

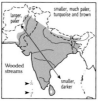

84.3. Indian Roller
Coracias b. benghalensis group

<31cm p.270
From European by rufescent face and breast with fine white stripes; boldly banded blue wings in flight.

84.2. 'Black-billed' Roller
Coracias benghalensis affinis

>31cm p.270
From widespread nominate group by smoky-brown unstriped face and breast, and darker upperparts.

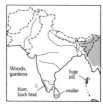

84.4. Dollarbird
Eurystomus orientalis

31cm p.270
Dark blue-green with short heavy red bill, eye-ring and legs; looks black in the field with very large blue-white carpal patch in flight. Juv. has dusky bill.

Hoopoe Upupidae

Unmistakable buffy-rufous bird, with long crest, long curved bill, and boldly pied mantle and wings. All plumages similar.

84.5. Common Hoopoe
Upupa e. epops group

31cm p.271
Buff, zebra-striped above with long curved bill, fan-like crest and short legs.

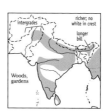

84.6. *Upupa epops ceylonensis*

Not illustrated. Resident *ceylonensis* darker, with shorter crest lacking white spots.

Trogons *Harpactes*

Sedentary, long-tailed and spectacularly coloured birds with short stout bills and short weak legs. Perch erect, but habits sluggish and movements slow; sexes and ages dissimilar. Note colour of head and upperparts.

84.7. Malabar Trogon
Harpactes fasciatus

31cm p.261
Smallish, with largely white undertail and bright rufous upperparts. Male has blackish hood, white breast-band and red underparts; female with brown hood and ochre underparts.

84.9. Ward's Trogon
Harpactes wardi

40cm p.262
Large and long-tailed, with dark grey head and breast with pale forehead and pinkish bill. Male has pink forehead, belly and outertail feathers, yellow in female.

84.8. Red-headed Trogon
Harpactes erythrocephalus

35cm p.261
Rather large with rufous upperparts, scarlet belly and large white tail-tips. Male has pinkish-crimson head and breast; rufous-brown in female.

84.10. Orange-breasted Trogon
Harpactes oreskios

p. 262
Possible; extreme SE (not illustrated). Small, yellow-bellied trogon with chestnut mantle and uppertail. Male has greener head and more orange belly, female's head greyer.

PLATE 85. KINGFISHERS I (A. Gilbert)

Kingfishers *Pelargopsis*

Very large with heavy red bills, buffy underparts and pale blue rumps. Note colour of head and upperparts.

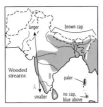

85.1. Stork-billed Kingfisher
Pelargopsis capensis

38cm   p.264
Dull blue above; on mainland with brownish cap and broad buff collar. Cap less distinct in Andamans, and lacking in Nicobars, where darker blue above.

85.2. Brown-winged Kingfisher
Pelargopsis amauroptera

36cm   p.264
Dark brown above with contrasting pale blue rump; head entirely buff-ochre.

Kingfishers *Halcyon*, *Todiramphus*

Mid-sized, stout-billed and brightly coloured.

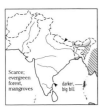

85.3. Ruddy Kingfisher
Halcyon coromanda

26cm   p.264
Entirely rufous with magenta tint above, bluish-white rump, and red bill and legs. Much larger than Rufous-backed Dwarf (Pl. 86), which has rufous rump.

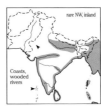

85.5. Black-capped Kingfisher
Halcyon pileata

30cm   p.265
Purplish-blue above with black cap and shoulder, white collar and red bill.

85.4. Collared Kingfisher
Todiramphus chloris

24cm   p.265
Greenish-blue above with white collar and underparts, and mostly black bill. White brow is short on mainland, reduced or lacking in Andamans (where cheeks blackish), and long and buffy in larger Nicobars *occipitalis*, forming band that meets at nape.

85.6. White-throated Kingfisher
Halcyon smyrnensis

28cm   p.264
Brilliant blue above with dark chestnut head, shoulder and underparts, white throat and breast, and red bill. Note large white carpal patch in flight.

Plate sponsored by Drue Heinz Fund.

PLATE 86. KINGFISHERS II (A. Gilbert)

Pied kingfishers *Ceryle*

Strongly patterned black and white, crested. Note size, head pattern, crest shape.

86.1. Himalayan Pied Kingfisher
Ceryle lugubris
41cm   p.265
Very large; banded black-and-white above with double-peaked shaggy crest, spotted malar and breast-band. Rufous wing-lining prominent in flight.

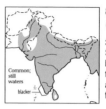

86.2. Lesser Pied Kingfisher
Ceryle rudis
31cm   p.266
Smaller than Himalayan and more boldly black-and-white, with white brow, tail-base and tip, and black crest, mask and breast-band (two in male, one in female).

Kingfishers *Alcedo*

Small and short-tailed kingfishers, black-and-turquoise above with white post-auricular spot, and rufous below. Note size, pattern of cheek and upperparts.

86.3. Blyth's Kingfisher
Alcedo hercules
22cm   p.262
Large version of Blue-eared, with crown more spangled than barred, and less distinct lore-spot; probably never has rufous on cheek. Female has extensive red on lower mandible.

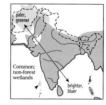

86.5. Common Kingfisher
Alcedo atthis
17cm   p.263
Paler above than congeners, with blue (rather than blackish) wings making less contrast with back-stripe, and rufous-orange cheek-patch (but see juv. Blue-eared). In open habitats rather than forest.

86.4. Blue-eared Kingfisher
Alcedo meninting
16cm   p.263
From Common by dark blue cheek and darker wings, making more contrast with pale dorsal streak and rump. Juv. has rufous cheek-patch like adult Common, but lacks blue moustache. Much smaller than similar Blyth's; female has duskier red lower mandible than female Blyth's.

Dwarf kingfishers *Ceyx*

Extremely small and short-tailed forest species; rufous with dark wings and eye-patch, white ear-patch and bright red bill.

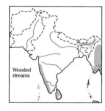

86.6. Black-backed Dwarf Kingfisher
Ceyx erithaca
13cm   p.263
Orange-rufous with blue-black mantle, wings and ear-spot.

86.7. Rufous-backed Dwarf Kingfisher
Ceyx rufidorsa   13cm   p.263

Vagrant; Sikkim. From Black-backed by rufous mantle.

PLATE 87. BEE-EATERS (J. Anderton)

Bee-eaters *Merops, Nyctyornis*

Agile aerial-feeding insectivores. Brightly coloured, with long decurved bills, narrow pointed wings and longish tails. Typical bee-eaters *Merops* are found in open country, and have long narrow central tail-streamers, black masks and red eyes; juveniles have washed-out pattern of adult. Note head pattern, colour of rump and tail.

87.1. **Chestnut-headed Bee-eater**
Merops leschenaulti

21cm   p.269

Small with notched, streamerless tail; green with chestnut crown and mantle, pale blue rump and yellow throat. Juv. duller and dilute.

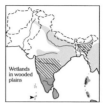

87.4. **Blue-tailed Bee-eater**
Merops philippinus

31cm   p.269

Golden-olive with bright blue rump and tail, bluish belly and vent, and broad chestnut throat-patch. Juv. is bluer above, and lacks pale brow of juv. Blue-cheeked.

87.2. **European Bee-eater**
Merops apiaster

27cm   p.269

Largely blue with rufous upperparts, golden scapulars and rump, white forehead and yellow throat; streamers rather short. Juv. is duller, with greener upperparts.

87.5. **Little Green Bee-eater**
Merops orientalis

21cm   p.268

Small and green, with coppery sheen to head and back, and bluish-green throat. Crown more rufous and throat greener in Myanmar. Juv. is uniform and dull.

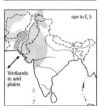

87.3. **Blue-cheeked Bee-eater**
Merops persicus

31cm   p.269

Greener overall than Blue-tailed with only slightly bluish rump and tail and less yellow on throat. Juv. is much duller, and bluer-green above.

87.6. **Blue-bearded Bee-eater**
Nyctyornis athertoni

36cm   p.268

Large with a square tail and pale eye; grass-green with turquoise forehead and throat-plumes and yellow-buff belly. Lacks black mask of typical bee-eaters.

Plate sponsored by Mr. Howard Phipps.

PLATE 88. HORNBILLS (A. Gilbert)

See Plate 89 for flight illustrations

Hornbills Bucerotidae

Large, often gregarious frugivores with huge bills often sporting casques, and vivid facial skin. Note head and bill pattern.

88.1. Narcondam Hornbill
Aceros narcondami

66cm   p.274

Miniature version of Wreathed. Male has rufous head and upper breast; juv. like dull male.

88.6. Rufous-necked Hornbill
Aceros nipalensis

122cm   p.274

Large, with black upperparts and white distal half of tail, yellow bill with black bars, red gular pouch, blue facial skin. Male otherwise rufous, female black.

88.2. Malabar Grey Hornbill
Ocyceros griseus

59cm   p.272

Grey above with uncasqued bill, cinnamon vent, and black tail with white tips.

88.7. Great Pied Hornbill
Buceros bicornis

130cm   p.273

Immense with large yellow bill and casque; black with buffy neck, white wing-bands, belly and tail, and black tail-band. Female has smaller casque without black trim, pale eye and red eye-ring.

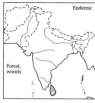

88.3. Ceylon Grey Hornbill
Ocyceros gingalensis

59cm   p.272

Resembles Malabar but darker and scaled above, whiter below, with outertail feathers usually all white. Male has yellow bill, blackish in female; juv. has all-pale bill.

88.8. Malabar Pied Hornbill
Anthracoceros coronatus

92cm   p.273

Mid-sized with yellow bill and huge black-topped pointed casque; black with white underparts and outertail feathers. Male has dark eye-skin.

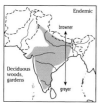

88.4. Indian Grey Hornbill
Ocyceros birostris

61cm   p.272

Brownish-grey with pointed casque, dark cheek, and white-tipped tail with black subterminal band.

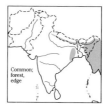

88.9. Oriental Pied Hornbill
Anthracoceros albirostris

89cm   p.273

From Malabar Pied by white-tipped black outertail feathers, and unpointed black-fronted casque. Female has dark reddish bill-base.

88.5. White-throated Brown Hornbill
Ptilolaemus austeni

76cm   p.274

Mid-sized; brown above, rufescent below, with low casque on pale yellowish bill and broad white corners on tail. Male has white cheek and throat.

88.10. Wreathed Hornbill
Aceros undulatus

110cm   p.274

Large, with pale corrugated bill and flat casque, and black stripe on gular pouch; black with white tail. Male has chestnut crown, white face and throat, and yellow pouch; female has blue pouch.

PLATE 89. HORNBILLS IN FLIGHT (J. Anderton) See Plate 88 for non-flight illustrations. Note size, wing and tail pattern.

Hornbills Bucerotidae

Most species show distinctive patterns of dark and light in flight, and have noisy flight that is often distinctive to species.

89.1. Ceylon Grey Hornbill
Ocyceros gingalensis p.272
Like Malabar Grey, but whiter below with outertail feathers usually all white.

89.2. Malabar Grey Hornbill
Ocyceros griseus p.272
Darker above and blacker-tailed than Indian Grey with small white carpal patch, grey wing-lining, and blackish undersides of primaries.

89.3. Indian Grey Hornbill
Ocyceros birostris p.272
Paler above than Ceylon and Malabar Grey, with white trailing edge on wings, black subterminal band and white tips on tail.

89.4. White-throated Brown Hornbill
Ptilolaemus austeni p.274
Dark above with broad white corners on tail and fine primary tips. Male has white cheek and throat; female lacks white tail-tips.

89.5. Malabar Pied Hornbill
Anthracoceros coronatus p.273
Mid-sized; black with white belly, outertail feathers and trailing edge of wing.

89.6. Oriental Pied Hornbill
Anthracoceros albirostris p.273
From similar Malabar Pied by white-tipped black outertail feathers.

89.7. Great Pied Hornbill
Buceros bicornis p.273
Immense, with black band across centre of white tail, broad white bar and trailing edge on black wings, and pale neck.

89.8. Rufous-necked Hornbill
Aceros nipalensis p.274
Large; black wings with broad white primary tips and black tail with white distal half. Male rufous otherwise, female black.

89.9. Wreathed Hornbill
Aceros undulatus p.274
Large; black with white tail, and no pale wing-tips. Male has white face and throat.

89.10. Narcondam Hornbill
Aceros narcondami p.274
Small and dark with entirely white tail.

PLATE 90. HONEYGUIDE AND BARBETS (J. Anderton)

Honeyguide Indicatoridae

Small, drab, dark superficially finch-like picid. Quiet and unobtrusive, usually found around rock bee nests.

90.1. Yellow-rumped Honeyguide
Indicator xanthonotus
15cm   p.280
Small; dusky olive with stubby bill, golden-yellow forehead, throat and rump. Compare with females of Scarlet and Crimson-browed finches (Pl. 169).

Barbets *Megalaima*

Clumsy-looking, green canopy-dwellers; among the primary hidden noise-makers in any forest. Note head pattern, size.

90.2. Coppersmith Barbet
Megalaima haemacephala
17cm   p.278
Blackish face with red forehead and breast-patch, yellow eye-ring and throat, and whitish below with dark streaks. Juv. lacks red and has pale patch below eye.

90.3. Ceylon Small Barbet
Megalaima rubricapillus
17cm   p.278
Bluish head with orange-yellow eye-ring and throat, and green underparts. Juv. like juv. Coppersmith but green below.

90.4. Malabar Barbet
Megalaima malabarica
17cm   p.278
Green with red face and throat, unstreaked green underparts. Juv. much like juv. Ceylon Small.

90.5. Yellow-fronted Barbet
Megalaima flavifrons
21cm   p.277
Mid-sized; yellow crown and moustache, blue face and throat, and dark scales on breast.

90.6. Blue-throated Barbet
Megalaima asiatica
23cm   p.278
Rather large with pale-based bill, blue hood with black crown and red forehead, nape and lateral breast-spot.

90.7. Blue-eared Barbet
Megalaima australis
17cm   p.278
Small with stubby black bill; blue face and throat, black forehead, yellow suborbital patch and red postocular streak and moustache spot. Juv. has head all green or nearly so.

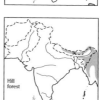

90.8. Golden-throated Barbet
Megalaima franklinii
23cm   p.277
Unique black mask, with red forehead, yellow crown and chin, grey cheek and throat. Juv. nearly lacks head markings.

90.9. White-cheeked Barbet
Megalaima viridis
23cm   p.277
Smallish with pale bill, dark brown head with white brow, cheek-patch and throat, and brownish-streaked buff breast.

90.10. Brown-headed Barbet
Megalaima zeylanica
27cm   p.277
From Lineated by browner head (especially throat) and breast, and contiguous bill and eye-patch (orange in breeding). Juv. has unstreaked buffy head and breast.

90.11. Lineated Barbet
Megalaima lineata
28cm   p.277
Large; with large yellow bill and (disjunct) eye-ring, whitish head with brown streaks, and usually unmarked throat. Juv. has less streaked head and breast.

90.12. Great Barbet
Megalaima virens
33cm   p.276
Large, with huge yellow bill, blackish-blue head, yellowish, streaked belly and red vent.

PLATE 91. WRYNECK, PICULETS AND SMALL PIED WOODPECKERS (J. Anderton)

Wryneck *Jynx*

Rather small, cryptically coloured, non-clinging woodpecker relative. Sexes and ages similar.

91.1. Eurasian Wryneck
Jynx torquilla
19cm p.280
Short-billed and long-tailed with nightjar-like plumage. Note narrow dark mask, barred tail.

Heart-spotted Woodpecker *Hemicircus*

Tiny, toy-like pied woodpecker, distinctive in region.

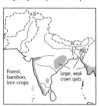

91.2. Heart-spotted Woodpecker
Hemicircus canente
16cm p.291
Small and rotund, with thin neck and short rounded tail; black above with long crest, buffy-white throat, and white tertials with heart-shaped black spots. Female has white forecrown.

Piculets

Tiny with very short tail. Sex/age differences minor.

91.3. Speckled Piculet
Picumnus innominatus
10cm p.281
Olive above and white below with dark mask and moustache, black spots on belly.

91.4. White-browed Piculet
Sasia ochracea
9cm p.281
Olive above with white eyebrow, rufous below.

Pied woodpeckers *Dendrocopos* (also on Plate 92)

All have dark upperparts with white bars, spots or patches, and pale underparts, usually with red on vent. Males have some red on head, lacking in female, but often more extensive in juvs. Note pattern of face and upperparts, especially tail.

91.5. Indian Pygmy Woodpecker
Dendrocopos nanus
13cm p.281
Very small; blackish-brown above with brown crown, whitish eye and barred tail. Blackish above in Sri Lanka, whitish and unstreaked below.

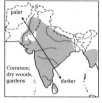

91.6. Yellow-fronted Pied Woodpecker
Dendrocopos mahrattensis
18cm p.283
Spotted above with yellow forecrown and barred tail, and washily streaked below with whitish vent.

91.7. Brown-fronted Pied Woodpecker
Dendrocopos auriceps
20cm p.282
More heavily streaked below than Yellow-fronted with brown forecrown, barred upperparts, unbarred tail, and red vent.

91.8. Grey-capped Pygmy Woodpecker
Dendrocopos canicapillus
14cm p.282
Where range overlaps Indian Pygmy, from that species by unbarred tail and grey crown. In E Himalayas, nape red, and in S Assam hills, uppertail banded.

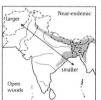

91.9. Fulvous-breasted Pied Woodpecker
Dendrocopos macei
19cm p.282
Small; buffy face and underparts with lightly spotted breast. Male has dull red crown.

91.10. Spot-breasted Pied Woodpecker
Dendrocopos analis
<19cm p.282
Like Fulvous-breasted, but whiter-faced with heavily spotted breast and barred tail. Female has rather brownish crown.

91.11. Stripe-breasted Pied Woodpecker
Dendrocopos atratus
21cm p.283
From Fulvous-breasted by whiter face and heavily streaked underparts.

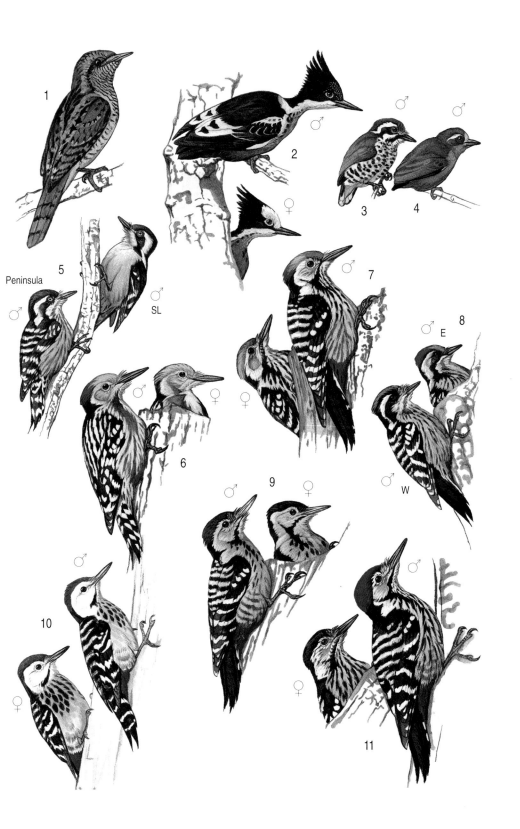

PLATE 92. LARGER PIED WOODPECKERS (J. Anderton)

Pied woodpeckers *Dendrocopos, Picoides* (continued from Plate 91)

Most of the pied woodpeckers shown here have black mantle with white wing- or scapular patches.

Pines, mixed hill forest

92.1. Rufous-bellied Woodpecker
Hypopicus hyperythrus

20cm p.285
Barred black-and-white above, solid rufous below. Female has black crown spotted with white; juv. (not illustrated) very different, mottled brown below.

Broadleaf forest

92.2. Crimson-breasted Pied Woodpecker
Dendrocopos cathpharius

18cm p.283
From much larger Darjeeling by smaller bill, more red on neck and breast, little on vent; less white on wing coverts. Male has unique red patch on nape extending down behind cheek.

Broadleaf forest

92.3. Darjeeling Pied Woodpecker
Dendrocopos darjellensis

25cm p.283
Large; fulvous below with black streaks and yellow-orange side of neck. Male has red nape-patch but orange rather than red (as in smaller Crimson-breasted) on side of neck.

Montane forest

92.4. Eurasian Three-toed Woodpecker
Picoides tridactylus

23cm p.285
Possible; N Arunachal. Smallish; largely black with buffy brow, throat and streaks on breast, and a broad white patch on lower back. Male has yellow crown, black in female.

Montane forest

92.5. Himalayan Pied Woodpecker
Dendrocopos himalayensis

25cm p.284
Large; lacks strong necklace; smallish pale cheek-patch and large pale neck-patch are separated by vertical black bar. Male has red crown and black nape. Smoky-buff below in E, paler in W.

Hill forest

92.6. Great Spotted Woodpecker
Dendrocopos major

24cm p.284
Large with large cheek-patch and small neck-patch (reverse of Himalayan), and broad black necklace on smoky-brown underparts. Male has red nape.

92.7. Syrian Woodpecker
Dendrocopos syriacus 23cm p.284

Possible; extreme W Afghanistan/Pakistan. Lacks cheek-bar of Himalayan and White-winged; from Sind Pied by blackish streaks on flank, and rather pale red vent. Male has red nape.

Riparian scrub, groves

92.8. White-winged Pied Woodpecker
Dendrocopos leucopterus

24cm p.284
Extensively white wing; very broad black moustache and cheek-bar with small pale neck-patch. Male has red nape and black crown.

Open woods, scrub, gardens

92.9. Sind Pied Woodpecker
Dendrocopos assimilis

22cm p.284
Clean black-and-white, with no black cheek-bar on white face. Male has red crown.

WC Himalayas
juv

PLATE 93. GREEN AND BROWN WOODPECKERS (J. Anderton)

Green woodpeckers *Picus*

Mid-sized woodpeckers with olive upperparts. The yellownapes have striking yellow mane-like crests; the rest are uncrested, with males having red in crown lacking in female. Note pattern of head and underparts.

93.1. Scaly-bellied Woodpecker
Picus squamatus

35cm   p.288
Larger than Streak-throated with plain throat and breast, and more heavily scaled belly. Much paler in NW, with weaker face pattern and scaling on belly.

93.2. Streak-throated Woodpecker
Picus xanthopygaeus

29cm   p.287
Mid-sized with spotty black moustache, weak streaks on throat, close but weak scales on belly, and lightly barred dark primaries and tail. Male has red crown.

93.3. Streak-breasted Woodpecker
Picus viridanus

29cm   p.287
From Streak-throated by bolder black moustache, nearly unstreaked dull brown throat and breast, and paler moustachial streak; note also dark red eye. Sole regional specimen previously treated as Laced *P. vittatus*, but that species now rejected (Appendix 2).

93.4. Greater Yellownape
Picus flavinucha

33cm   p.287
From smaller Lesser by blackish breast and black bars on chestnut flight feathers. Male lacks red on head and has yellow throat, chestnut in female.

93.5. Lesser Yellownape
Picus c. chlorolophus

27cm   p.286
Smaller-billed than larger Greater with plain olive breast, white bars on belly, and almost unbarred flight feathers. Male has red moustache. In N, male has olive crown and red brow.

93.6. *Picus chlorolophus chlorigaster/ wellsi*

In W Ghats (*chlorigaster*) and Sri Lanka (*wellsi*) smaller and much duskier overall, with little yellow in short crest, little or no white on face; red crown in male.

93.7. Grey-faced Woodpecker
Picus canus

32cm   p.288
Diagnostic grey head with black nape and unmarked underparts.

Brownish woodpeckers *Micropternus, Blythipicus, Gecinulus*

Miscellaneous mid-sized, mostly brown species.

93.8. Rufous Woodpecker
Micropternus brachyurus

25cm   p.285
Smaller, chunky with short bill and lax rounded crest; entirely rufous and finely barred.

93.9. Bay Woodpecker
Blythipicus pyrrhotis

27cm   p.290
Larger, with long yellowish bill; chestnut with pale head and black-banded upperparts.

93.10. Pale-headed Woodpecker
Gecinulus grantia

25cm   p.290
Smaller; maroon above with buffy head, and dusky olive below.

PLATE 94. FLAMEBACKS AND LARGE WOODPECKERS (J. Anderton)

Flamebacks or Goldenbacks *Dinopium, Chrysocolaptes*

Flameback (goldenback) woodpeckers comprise two genera: *Dinopium* are smaller and rather short-necked, with dark eyes, short bills and black hindnecks. *Chrysocolaptes* are larger and more gangly, with long stout bills, white hindnecks, and split moustache streaks. Females of both groups have black or partially black crests marked with white (except in female White-naped). Note rump colour, hindneck and moustache stripe patterns.

94.1. Common Flameback
Dinopium javanense

28cm p.289
Diagnostic combination of red rump and single moustachial streak. Female has white streaks on black crest.

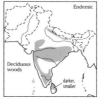

94.4. White-naped Flameback
Chrysocolaptes festivus

29cm p.290
Black above with white upper mantle, golden wings. Female has unique yellow crest.

94.2. Himalayan Flameback
Dinopium shorii

31cm p.288
From Common by split moustache and fulvous breast; from very similar Greater by black hindneck. Female has white streaks on black crest.

94.5. Greater Flameback
Chrysocolaptes lucidus

33cm p.289
Diagnostic combination of white hindneck and split moustache streak. Female has white spots on black crest.

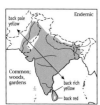

94.3. Black-rumped Flameback
Dinopium benghalense

29cm p.289
Diagnostic black rump and throat; note also black-and-white shoulder and primaries. Female has black forecrown with white spots. In S Sri Lanka, crimson above; smaller than similar sympatric Crimson-backed, and told by shorter bill, black throat and clearer white brow.

94.6. Crimson-backed Flameback
Chrysocolaptes stricklandi

33cm p.290
Like Greater but dark red above with paler bill and weaker pale brow.

Large woodpeckers *Dryocopus, Mulleripicus*

Dryocopus woodpeckers are large and mainly black with pale eyes and red crests. Females have black forecrowns, no red moustache. *Mulleripicus* is mostly grey; slight sex/age variation.

94.7. Black Woodpecker
Dryocopus martius

46cm p.286
Hypothetical; W Bengal; possible N Arunachal. Solid black with whitish bill. Male has red crown but black moustache; female has red hindcrown.

94.9. Andaman Woodpecker
Dryocopus hodgei

38cm p.286
Like larger White-bellied, but body solid black.

94.8. White-bellied Woodpecker
Dryocopus javensis

48cm p.285
Very large with white belly and rump, and long black bill. Male has red moustache and crown; female has red hindcrown. In Myanmar, more white on head, wing, vent.

94.10. Great Slaty Woodpecker
Mulleripicus pulverulentus

51cm p.291
Huge and long-necked with very long pale bill; slate-grey with yellow-buff throat. Male has pinkish moustache spot.

PLATE 95. BROADBILLS AND PITTAS (J. Anderton)

Broadbills *Serilophus, Psarisomus*

Big-headed forest birds with very broad bills and short weak legs.

95.1. Long-tailed Broadbill
Psarisomus dalhousiae

27cm   p.293

Bright green with black cap, yellow face, throat and collar, and long pointed blue tail.

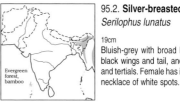

95.2. Silver-breasted Broadbill
Serilophus lunatus

19cm   p.293

Bluish-grey with broad black eyeline, black wings and tail, and rufous rump and tertials. Female has inconspicuous necklace of white spots.

Pittas *Pitta, Anthocincla*

Portly terrestrial birds with long legs, stubby tails and distinctive loud voices. Secretive, keeping to thick cover and generally easiest to see at dawn or dusk. Sexes similar or nearly so. Juveniles duller, with crown more or less spotted. Note head pattern, colour of throat and breast.

95.3. Hooded Pitta
Pitta sordida

19cm   p.294

Mid-sized; green with black head, chestnut crown, and red lower belly and vent. Juv. is brown below with pale throat and blackish chin; crown is brown and scaly.

95.6. Blue Pitta
Pitta cyanea

23cm   p.294

Mid-sized, with black mask and malar and red nape on brownish head, blue rump and tail, and barred whitish underparts. Male has blue mantle, olive in female. Juv. is buffy below with diffuse scales.

95.4. Mangrove Pitta
Pitta megarhyncha

23cm   p.296

Very like Indian, but bill much larger; brownish crown lacks black coronal stripe, no white eye-ring on black mask, and brow is duskier. Shoulder and rump are more purplish-blue, and underparts rather darker buff, less yellowish than Indian.

95.7. Blue-naped Pitta
Pitta nipalensis

25cm   p.294

Large; green above and ochre below. Male has blue nape.

95.8. Eared Pitta
Anthocincla phayrei   22cm   p.294

Hypothetical; Bangladesh. Entirely buff-brown with blackish mask, and scaly crown and wing-coverts; see text.

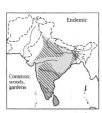

95.5. Indian Pitta
Pitta brachyura

19cm   p.295

Smallish; green above with black mask and coronal stripe on buff crown, white brow and throat, turquoise shoulder and rump and buff below with red belly and vent; adult's back sometimes black-streaked. Juv. browner below with less red, and a scaly crown.

PLATE 96. BUSH-, SKY- AND CRESTED LARKS (B. Zetterström)

Bushlarks *Mirafra*

Chunky, slightly crested larks with moderately heavy bills and some rufous in primaries. Juveniles scaled whitish above. Note tail length, breast-streaking, colour of outer rectrices, bill size.

96.1. Singing Bushlark
Mirafra cantillans

15cm p.296
Pale, relatively long-tailed and short-billed, with barely marked cheek and breast; note white outer rectrix, little rufous in wing.

96.3. Jerdon's Bushlark
Mirafra affinis

14cm p.297
From Indian by longer bill and legs, shorter tail and narrower brow buffy in front of eye; buffier below with heavier streaks on breast.

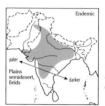

96.2. Indian Bushlark
Mirafra erythroptera

14cm p.297
Pale with heavily streaked crown and upperparts, clear rufous wing-panel, and large spots on breast; cheek-patch completely enclosed by whitish border.

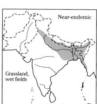

96.4. Bengal Bushlark
Mirafra assamica

14cm p.297
Chunky, with longish bill and short tail; dusky and weakly streaked above, with greyish-buff supercilium and underparts. Streaks on breast less defined than on Jerdon's or Indian, rufous wingpatch very prominent.

Skylarks and Woodlark *Alauda, Lullula*

Crests shorter and bushier than crested larks, with whitish outertail feathers; similar where ranges overlap. Note colour of cheek and wing-trim.

Crested larks *Galerida*

Rather nondescript with long, spiky crests, buffy outertail feathers. Note size and bill shape, degree of buffiness and streaking.

96.5. Eurasian Skylark
Alauda arvensis

18cm p.307
Larger than Oriental, with longer tail and wings and smaller bill; usually paler, with pale buffy edges to flight feathers and broader white trailing edge to secondaries.

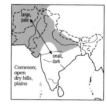

96.8. Crested Lark
Galerida cristata

18cm p.305
Large, long-billed with pale, weakly streaked upperparts. Sandy above in W Pakistan, greyer to E.

96.6. Oriental Skylark
Alauda gulgula

16cm p.307
Smaller than Eurasian, with warmer brown auriculars and richer rufous in flight feathers.

96.9. Malabar Lark
Galerida malabarica

15cm p.305
Darker and more heavily streaked above than Crested; from Sykes's by streaked rump, dark rear auriculars and whitish belly.

96.7. Woodlark
Lullula arborea 14cm p.307

Hypothetical; SW Afghanistan. Chunky, with very short, white-cornered tail; note bright brown cheek enclosed by dark eyeline, and prominent white-tipped black carpal patch.

96.10. Sykes's Lark
Galerida deva

13cm p.306
Small and shortish-billed, with pinkish-buff brow and underparts, pale rear auriculars and unstreaked buffy rump; less streaked on central breast than Malabar.

PLATE 97. FINCH-, CALANDRA, HORNED AND HOOPOE LARKS (B. Zetterström)

Finch-larks or Sparrow-larks *Eremopterix*

Small, stout-billed with strongly curved culmens. Males have black brow and underparts; females are dull and sparrowlike. Juveniles scaled whitish above. Note crown colour and streaking, underpart pattern.

97.1. Ashy-crowned Finch-lark
Eremopterix griseus

13cm   p.299
Male has grey crown and blackish winglining. Female from female Black-crowned by buffier brow, darker cheek, weaker streaks on crown and diffuse streaks on entire underparts.

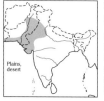

97.2. Black-crowned Finch-lark
Eremopterix nigriceps

13cm   p.299
Male has black crown and grey winglining. Female paler than female Ashy-crowned, with sharper streaks on crown and on buffier breast, and unstreaked belly.

Calandra larks *Melanocorypha*

Large, heavy-set, thick-billed larks with black patch on upper breast. Note size, bill length, face pattern, and amount of white in tail.

Miscellaneous larks *Alaemon, Eremophila*

Large, heavy-set, thick-billed larks with black patch on upper breast. Note size, bill length, face pattern, and amount of white in tail.

97.3. Bimaculated Lark
Melanocorypha bimaculata

16cm   p.301
Shorter-tailed than Calandra, with darker streaks on crown and mantle, stronger black eyeline (especially behind eye), clearer buff brow and warmer brown cheek. In flight note white tail-tips (not sides); wings lack Calandra's strong white trailing edge.

97.6. Greater Hoopoe Lark
Alaemon alaudipes

23cm   p.309
Large and slender with a long curved bill and long pale legs; pale with bold blackish eyeline, whisker mark and dark spots on breast. In flight note black-and-white-banded wings.

97.4. Calandra Lark
Melanocorypha calandra

18.5cm   p.301
Larger, longer-winged and -tailed than Bimaculated, with plainer face and upperparts and shorter, stouter bill. In flight, note white trailing edge on dark wing and white outertail feathers.

97.7. Horned Lark
Eremophila alpestris

20cm   p.308
Pale and little-streaked with a small dark bill, and black forecrown, mask and breast-patch. In fresh plumage, black patches partly obscured. Juv. dark with white spots above, and mottled dusky breast.

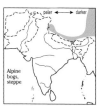

97.5. Tibetan Lark
Melanocorypha maxima

21cm   p.302
Very large with long heavy bill and strong black legs; heavily streaked above with grey nape, small black chest-patches, and no dark eyeline.

PLATE 98. DESERT LARKS AND SHORT-TOED LARKS (B. Zetterström)

Desert larks Ammomanes

Rather small and thick-billed larks with unstreaked backs and rufescent rumps. Juveniles have wide buffy edges on wings. Note tail pattern, breast-streaking.

98.1. Rufous-tailed Lark
Ammomanes phoenicura

16cm   p.299
Dusky with rufous belly, rump and tail and broad blackish terminal band; hardly streaked except on breast.

98.2. Desert Lark
Ammomanes deserti

16cm   p.300
Variably coloured, with dusky, rufescent-based tail lacking clear-cut pattern; less rufous on crown and wings than Bar-tailed, with more streaked breast.

98.3. Bar-tailed Lark
Ammomanes cinctura

16cm   p.301
Very pale and plain with clear-cut black terminal band on pale rufous tail.

Short-toed larks Calandrella

Confusing, small and plain with short conical bills and white outertail feathers. Juveniles heavily speckled whitish above. Note facial pattern, bill size and colour, primary projection, and presence of chest-patch, streaks on breast.

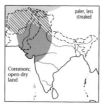

98.4. Greater Short-toed Lark
Calandrella brachydactyla longipennis

15cm   p.303
Warm brown and streaked above with pinkish bill, little primary projection; note broad pale lores (without dark line of Hume's), brow and suborbital patch, small dark patch and light streaks on breast.

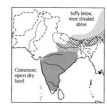

98.5. *Calandrella brachydactyla dukhunensis*

Eastern race *dukhunensis* is tawny-breasted and less grey overall than western *longipennis*.

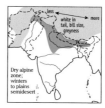

98.6. Hume's Short-toed Lark
Calandrella acutirostris

14cm   p.303
From congeners by blackish line on lores and blackish ridge on yellow bill; note dark cheek, weakly streaked crown and unstreaked breast. No primary projection.

98.7. Lesser Short-toed Lark
Calandrella rufescens

15cm   p.304
Paler and heavier-billed than Greater, with clearer fine streaks on breast, weaker streaks on upperparts, and substantial primary projection.

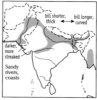

98.8. Sand Lark
Calandrella raytal

13cm   p.305
Small and stocky with short tail; most like Lesser Short-toed, but finer-billed with less crisp streaks on breast (but stronger than on Hume's and Greater), and shorter primary projection. In NW, *adamsi* has longish smudgy streaks (*vs* short crisp streaks of Lesser); NE race has longer, curved bill.

PLATE 99. MARTINS (J. Anderton)

Swallows and martins are distinguished from the superficially similar swifts (Plates 82–83) by their graceful, less nervous flight, and broader wings; they prefer high, exposed perches. Note tail shape, rump colour. Immatures are duller than adults, generally with stubbier tails.

Sand-martins *Riparia*

Small and brown with white belly; no white tail-spots. Found near water; flight rapid. Note presence of collar, shape of tail.

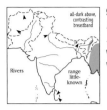

99.1. Common Sand-martin
Riparia riparia

13cm   p.309
Well-defined dark collar separates white throat from white belly. See text.

99.3. Pale Sand-martin
Riparia diluta indica

13cm   p.310
Not illustrated. White throat above diffuse breast-band; tail with shallow notch. Separation from Common not straightforward owing to poorly understood racial variation; see text.

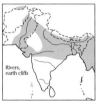

99.2. Grey-throated Sand-martin
Riparia chinensis

12cm   p.310
Silvery-grey throat and breast, and slightly pale rump; no breast-band. Tail nearly square. Very like NW race *diluta* of Pale; see text.

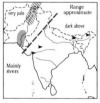

99.4. *Riparia d. diluta*

NW wintering *diluta* (above dashed line on map) is very pale and easily mistaken for Grey-throated; scattered records of darker *tibetana* (which recalls Common but has diffuse breastband) elsewhere (below dashed line).

Crag-martins *Ptyonoprogne*

Larger than sand-martins with white spots on rounded tails prominent in banking flight. Found near cliffs and villages up into the hills. Note colour of rump, colour and pattern of underparts (especially throat and vent).

House-martins *Delichon*

Size of sand-martins, but look black above with white rump and underparts. Primarily Himalayan; around settlements, cliffs and river valleys. Note colour of throat and vent, shape of tail.

99.5. Pale Crag-martin
Ptyonoprogne obsoleta

13cm   p.311
Very pale with whitish belly, wing-lining and rump.

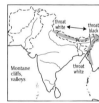

99.8. Nepal House-martin
Delichon nipalense

13cm   p.313
Small with black chin (sometimes very limited), vent and wing-lining, and nearly square tail.

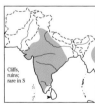

99.6. Dusky Crag-martin
Ptyonoprogne concolor

13cm   p.311
Entirely dark with streaked rufous underparts; often appears blackish in field.

99.9. Asian House-martin
Delichon dasypus

14cm   p.314
Smaller than Northern with sullied breast and rump (latter not extending as high on back), grey-brown wing-lining, and only moderately notched tail.

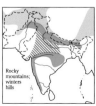

99.7. Eurasian Crag-martin
Ptyonoprogne rupestris

14cm   p.311
Large with pale breast and dark wing-lining, belly and vent; tail has larger spots than other crag-martins.

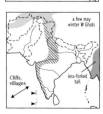

99.10. Northern House-martin
Delichon urbicum

15cm   p.314
Bright white below and on rump, with pale grey wing-lining, and sharply forked tail. Juv. much duller.

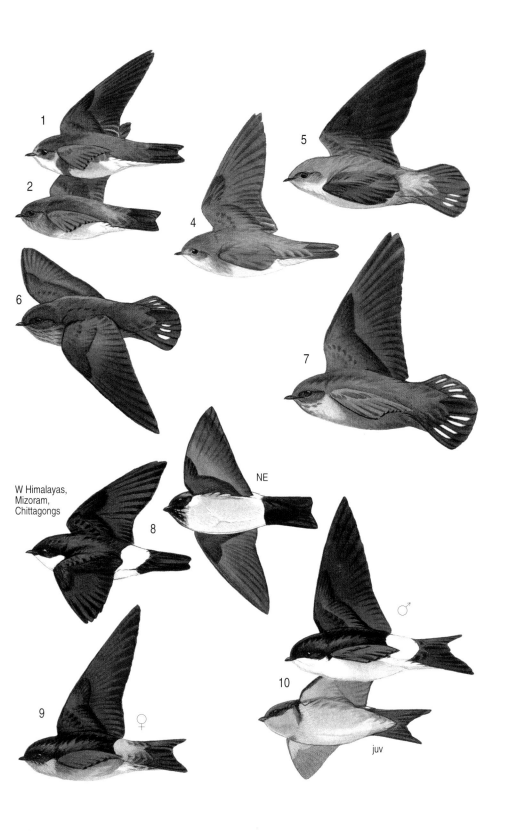

PLATE 100. SWALLOWS (J. Anderton)

100.1. Crested Treeswift
Hemiprocne coronata   see Pl. 82, p.361

Superficially like some fork-tailed swallows, but note large forehead crest; perches upright unlike other swifts.

Swallows *Hirundo*

Dark and glossy above, and pale below with rusty plumage on head and/or foreparts. Tails on some species deeply forked.

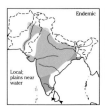

100.2. Streak-throated Swallow
Hirundo fluvicola

12cm   p.313
Tiny with very short notched tail, rufous crown, whitish rump and white underparts with streaked throat and breast.

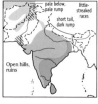

100.7. Red-rumped Swallow
Hirundo daurica (part)

19cm   p.312
Mostly resident, little-streaked races.

100.3. Hill Swallow
Hirundo domicola

13cm   p.312
Very like House, with greener gloss above; see text.

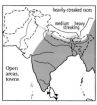

100.8. *Hirundo daurica* (part)

Mostly wintering, well-streaked races.

Dark above with rufous face, hindcollar and rump, whitish below with fine streaks and black vent; no white spots on deeply forked, unspotted tail. Streaks below much reduced on subadult, and in NW *rufula*, which also

100.4. House Swallow
Hirundo tahitica

13cm   p.312
Like Barn, but smaller with short notched tail, and dingy below with no breast-band.

whitish on rump; heavier on wintering races nominate and *japonica*, which approach Striated.

100.9. Ceylon Swallow
Hirundo hyperythra

18cm   p.313
Darker rufous than Red-rumped, with no hindcollar; shorter tail.

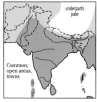

100.5. Barn Swallow
Hirundo rustica (part)

18cm   p.311
Dark upperparts and head with chestnut forehead and throat, unstreaked pale belly and vent, and white spots on deeply forked tail.

100.10. Striated Swallow
Hirundo striolata

19cm   p.313
From Red-rumped (with difficulty, especially in NE in winter) by adult's heavier streaks below, darker rufous face and rump. See text.

100.6. *Hirundo rustica tytleri*

NE migrant race *tytleri* is dark chestnut below with incomplete breast-band, usually very long tail streamers.

100.11. Wire-tailed Swallow
Hirundo smithii

14cm   p.312
Tail looks very short; long fine streamers barely visible. Entirely white below, and dark above with rufous cap.

PLATE 101. WAGTAILS (J. Anderton)

Wagtails Motacilla, Dendronanthus

Strongly patterned, longish-tailed birds of open habitats, usually near water. Some species extremely variable in plumage. *Dendronanthus* is atypical in morphology and habitat preference. Note colour of upperparts (especially rump), face and wing pattern.

101.1. Western Yellow Wagtail
Motacilla flava leucocephala

17cm   p.316
Pale little-patterned head, grey in female.

101.2. *Motacilla flava feldegg*

Male has black crown, olive-tinged in winter; female highly variable but distinctive (see text)

101.3. *Motacilla flava thunbergi*

Male's crown browner, throat whiter in winter; can have hint of short white brow.

101.4. *Motacilla flava beema*

Both sexes pale grey-headed with strong white supercilium. All races relatively short-tailed, with two wing-bars, olive-green above and (usually) yellow below. See text. Juv. of any race may be buffy or white below, lacking yellow, and lacking Citrine's complete pale border around dark cheek-patch.

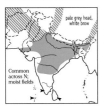

101.5. **Grey Wagtail**
Motacilla cinerea

17cm   p.317
Slim with long tail; grey above with yellowish rump and vent, pale eyebrow, plain wing. Breeding birds yellow below; male has black throat, white submoustachial, female usually has all-white throat. Non-breeders buffy-whitish below.

101.6. **Citrine Wagtail**
Motacilla c. citreola

17cm   p.316
Rather large with broad white wing-bars; lacks olive above. Breeding male has yellow head and underparts; female has greyish crown and cheek, latter enclosed by yellowish brow and neck-collar. Pale first-winter may lack yellow, but face pattern like female's. Juv. browner above, with broken dark necklace on buffy underparts (see Forest).

101.7. *Motacilla citreola calcarata*

Breeding male *calcarata* has black upperparts; grey-backed much like nominate in winter.

101.8. **Forest Wagtail**
Dendronanthus indicus

17cm   p.315
Portly, pipit-like and relatively short-tailed; olive-brown above with white brow and wing-bars, and white below with diagnostic bold black double necklace.

101.9. **White Wagtail**
Motacilla alba dukhunensis

18cm   p.315
All races white below and grey or black above with white forehead, largely white wing, and black hindcrown and breast; much racial variation. Throat black in most breeding plumages, white in all non-breeders. First-winter can be suffused yellowish below; forehead grey in juv.

101.10. *Motacilla alba personata*

101.11. *Motacilla alba ocularis*

Only form with thin black eyeline.

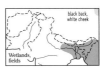

101.12. *Motacilla alba leucopsis*

Whole throat white in all plumages. Race *baicalensis* similar, with lower half throat black; in NE but much rarer.

101.13. *Motacilla alba alboides*

'Black-backed' Wagtail racial group *alboides* has mantle black in breeding plumage, dark grey in non-breeding.

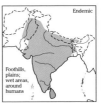

101.14. **White-browed Wagtail**
Motacilla maderaspatensis

21cm   p.315
Large and long-tailed; black above with white eyebrow and black forehead, white belly and wing-panel. Juv. browner but similar.

PLATE 102. OLIVE PIPITS (T. Schultz)

Pipits *Anthus* (continued on Plate 103)

Brownish and usually streaked slim birds of open habitats, grassland and agriculture, typically with thin bills, pale outertail feathers and long hind-claws. Breeding birds give diagnostic song-flights, but non-breeders are often skulking and difficult to identify except by voice. See especially larks (Pl. 96–98).

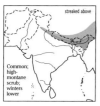

102.1. Olive-backed Pipit
Anthus h. hodgsoni

15cm   p.320
Greenish above with strong brow (buffy before eye, white after), pale spot behind cheek, black malar, and buffy on breast with bold black streaks. Nominate moderately streaked above.

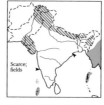

102.6. Red-throated Pipit
Anthus cervinus

15cm   p.320
Small and heavily streaked but 'clean'. Adult has rufous eyebrow, face and throat; first-winter has warm brown cheek, and heavy moustachial and malar streaks.

102.2. *Anthus hodgsoni yunnanensis*

Upperparts streaking minimal in *yunnanensis*.

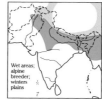

102.7. Rosy Pipit
Anthus roseatus

15cm   p.321
Olive above. Breeding plumage has pale pink eyebrow, throat and breast contrasting with blackish cheek. Non-breeding has heavily streaked mantle and breast, unstreaked rump, dark cheek bordered by pale buff.

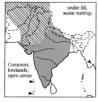

102.3. Tree Pipit
Anthus t. trivialis

15cm   p.319
Browner than Olive-backed, more streaked above and less streaked on breast; larger than Meadow, and less streaked above, with clearer eyebrow and dark eyeline.

102.8. Buff-bellied Pipit
Anthus rubescens

15cm   p.321
Darker than similar Water, greyer above with stronger moustachial stripe, and usually paler legs. Extensively buff and lightly marked below in breeding plumage, whiter and very heavily streaked in non-breeding.

102.4. *Anthus trivialis haringtoni*

NW breeding *haringtoni* is generally darker than nominate and more heavily streaked with a stouter, mostly dark bill.

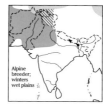

102.9. Water Pipit
Anthus spinoletta

15cm   p.321
Longish thin bill and darkish legs; greyish and scarcely streaked above. Pinkish-buff below in breeding plumage, and streaked only on flanks; whiter and vaguely streaked below in non-breeding.

102.5. Meadow Pipit
Anthus pratensis

15cm   p.320
From bulkier Tree by plainer face with weak eyebrow and no dark eyeline; more heavily streaked above.

PLATE 103. OCHRE PIPITS (T. Schultz)

Pipits *Anthus* (continued from Plate 102)

103.1. Richard's Pipit
Anthus richardi

18cm   p.317

Large, long-necked and long-legged, with distinctive upright stance; dark above with broad black streaks. Less uniform below than Blyth's, with whiter throat and mid-belly and more orange-buff flanks; centres of median coverts more V-shaped in adult. First-winter resembles Blyth's.

103.2. Paddyfield Pipit
Anthus rufulus

15cm   p.318

Smaller and dumpier than Richard's and Blyth's with shorter tail; also from Richard's by shorter bill and legs. From Blyth's by slightly darker lores, more crisply streaked crown, less distinctly streaked mantle, and whitish (not buffy) throat and belly. Very pale in NW; dark like Nilgiri in SW.

103.3. Tawny Pipit
Anthus campestris

15cm   p.318

Long-tailed, very pale sandy and nearly unstreaked, with pale bill; pale face and brow with fine black loral, whisker and malar lines. Juv. has lightly streaked and spotted breast. See Long-billed.

103.4. Blyth's Pipit
Anthus godlewskii

17cm   p.319

Between Richard's and Paddyfield (which see) in size and shape; distinctly and heavily streaked blackish and pale straw above; fairly uniform buff below. Squarish dark centres to median coverts in adult but not first-winter.

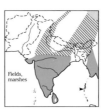

103.5. Nilgiri Pipit
Anthus nilghiriensis

17cm   p.321

Dark, richly coloured and heavily streaked, with dark lores, buffy eyebrow, and dark streaks on buffy breast and flanks. No white in plumage, and throat lacks malar of similar species.

103.6. Long-billed Pipit
Anthus similis

20cm   p.319

Large, robust, almost thrush-like, with long dark bill, long tail and short hindclaw. Nearly unstreaked dull pale brown with buff outer rectrices in NW; dark, heavily streaked in SW.

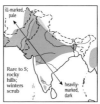

103.7. Upland Pipit
Anthus sylvanus

17cm   p.322

Rotund, thick-billed, and rather lark-like. Heavily streaked above with white eyebrow and eye-ring, no wing-bars; has long fine dark streaks below. Warm-toned in fresh plumage, duller when worn.

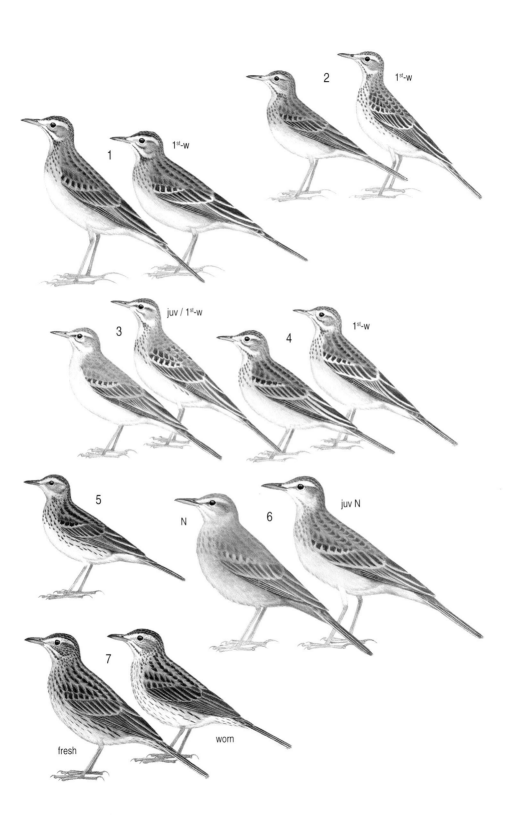

PLATE 104. WOODSHRIKES, CUCKOOSHRIKES AND WOOD-SWALLOWS (J. Anderton)

Woodswallows *Artamus*

Aerial feeders that gather on high exposed perches; heavier-set than swallows, with large silvery beaks and rounded tails. Note belly colour.

104.1. White-breasted Woodswallow
Artamus leucorynchus
18cm   p.593
Darker above than Ashy, with white underparts and rump; lacks white tail-tip.

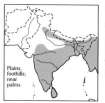

104.2. Ashy Woodswallow
Artamus fuscus
19cm   p.593
Pinkish-grey belly and white tail-tip; white trim on uppertail-coverts but not rump.

Woodshrikes *Tephrodornis*

Heavy-billed arboreal birds recalling shrikes, but with shorter tails and white on rump. Note size of rump-patch, presence of supercilium.

104.3. Common Woodshrike
Tephrodornis pondicerianus
16cm   p.330
Dull brown overall but for whitish brow, white sides to blackish tail; little white on rump. Juv. has white-spotted crown.

104.5. Large Woodshrike
Tephrodornis gularis
23cm   p.330
Large with large white rump-patch and solid brown tail; no pale brow. Female duller with paler bill and darker eye.

104.4. Ceylon Woodshrike
Tephrodornis affinis
18cm   p.331
Shorter-tailed than Common. Male has pale submoustachial, weaker eyebrow, bigger white rump-patch. Female duller and browner.

104.6. Malabar Woodshrike
Tephrodornis sylvicola
23cm   p.330
Both sexes darker overall than Large, with pale eyes.

Cuckooshrikes Campephagidae (continued on Plate 105)

Slim, upright-perching forest birds with hooked bills and short legs; genera on this plate mainly black, white and/or grey.

104.7. Pied Triller
Lalage nigra
18cm   p.324
White brow, wing-trim and tail-corners. Male black above; female greyer, streaked below.

104.10. Black-headed Cuckooshrike
Coracina melanoptera
20cm   p.323
Shortish white-tipped black tail, whitish belly. Male slaty with black hood; female browner with white brow, and barred throat and breast.

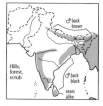

104.8. Pied Flycatcher-shrike
Hemipus picatus
14cm   p.328
Very small and longish-tailed; dark (black or brown) above with white wing-slash, tail-sides.

104.11. Andaman Cuckooshrike
Coracina dobsoni
26cm   p.323
Slaty above with red eye, longish untipped tail, heavy bars on white belly and vent. Male has grey throat and upper breast; female all barred below. First-winter similar, suffused rufous below.

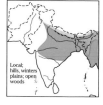

104.9. Black-winged Cuckooshrike
Coracina melaschistos
22cm   p.323
Solid slate-grey with white-tipped black tail; female usually pale on vent, finely barred below.

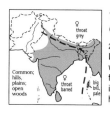

104.12. Large Cuckooshrike
Coracina macei
28cm   p.322
Large with heavy bill, dark mask; grey with white belly, tail-tips. S male and N female have barred belly. S female and imm. barred on throat, breast. Paler, heavier-billed in Andamans.

PLATE 105. MINIVETS (J. Anderton)

Minivets *Pericrocotus*

Rather small, active cuckooshrikes, mostly colourful and boldly patterned, especially males, but often difficult to identify to species. Subadult males often resemble females but with patchy plumage. Note wing and, in 'yellow' females, head patterns.

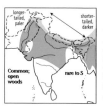

105.1. Long-tailed Minivet
Pericrocotus ethologus

18cm   p.326
Male cherry-red with double wing-slash. Female has single yellow wing-slash and small forehead-patch, pale throat and whitish cheek (but see text for S Assam hills).

105.2. Short-billed Minivet
Pericrocotus brevirostris

17cm   p.327
Both sexes with single wing-slash. Male blood-red; female with broadly yellow cheek and forehead.

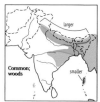

105.3. Scarlet Minivet
Pericrocotus speciosus

20cm   p.327
Large with heavy bill and tail, 'drops' on tertials. Male scarlet-red; female has broad golden-yellow forehead and face.

105.4. Orange Minivet
Pericrocotus flammeus

19cm   p.327
Smaller, shorter-tailed than Scarlet, with similar wing pattern. Male orange; female slatier-grey with yellow spectacles; much of cheek and forehead dark grey.

105.5. Grey-chinned Minivet
Pericrocotus solaris

17cm   p.326
Slaty-grey above with whitish throat. Male bright orange; female with dark forehead and cheek, yellow underparts.

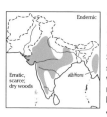

105.6. White-bellied Minivet
Pericrocotus erythropygius

15cm   p.326
Small, with white wing-slash and tail-sides, and orange rump. Male black with white belly and orange breast; female brownish-grey above with white brow and underparts. In Myanmar, Jerdon's *P. albifrons* has a white brow.

105.7. Small Minivet
Pericrocotus cinnamomeus

15cm   p.325
Small; grey with orange rump, wing-slash and tail-sides, and white belly. Male has black face and throat and orange breast; female with white spectacles and throat, and some yellow below.

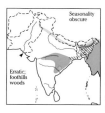

105.8. Rosy Minivet
Pericrocotus roseus

18cm   p.325
Brownish-grey with whitish throat and spectacles; single wing-slash; fine tertial-edges and tail-sides red in male, yellow in female. Male pink below and on rump; female yellowish below with dark rump.

105.9. Brown-rumped Minivet
Pericrocotus cantonensis   18cm   p.325

Hypothetical; Bangladesh. Brownish rump (not a clear patch) and sullied breast. Male like male Ashy but less clean-cut in pattern, with white shafts on tail. Female especially dull with small yellowish wing-slash (much smaller than in generally greyer female Rosy).

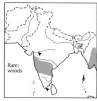

105.10. Ashy Minivet
Pericrocotus divaricatus

18cm   p.325
Cooler grey above than Brown-rumped, often with tiny whitish wing-patch, and whiter breast. Male has bolder head pattern with black hindcrown; female is greyish-crowned, with tertials fringed white.

PLATE 106. GREY AND OLIVE BULBULS (J. Anderton)

Bulbuls Pycnonotidae (continued on Plate 107)

Some of the most familiar Indian songbirds; most species are social and noisy, and frequent gardens and scrub as well as forest. The *Pycnonotus* species are rather short-billed and round-tailed; many are crested and boldly patterned, but some recall the *Hypsipetes* group (including *Iole* and *Hemixos*) which are generally plainer and longer-billed, most with longish, squared tails and ragged crests. Vocalisations of the latter group are usually wheezy and nasal, distinct from the musical chuckles of most *Pycnonotus*. *Alophoixus* may be gregarious, and recalls some babblers in behaviour.

106.1. Red-vented Bulbul
Pycnonotus cafer

20cm   p.338
Scaly brown with short-crested black hood, black tail with white tail-tips, white rump and red vent.

106.7. Ashy Bulbul
Hemixos flavala

20cm   p.342
Grey with green-gold wing-patch, white throat, black mask and brown ear-patch.

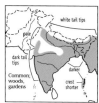

106.2. Red-whiskered Bulbul
Pycnonotus jocosus

20cm   p.337
Long slim crest on black crown, white cheek and underparts with red ear-tuft and vent; no white on rump. Broad breast-band and white tail-tips are variable.

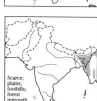

106.8. Olive Bulbul
Iole virescens

19cm   p.341
Small and drab; olive above with rufescent crown, wings and tail, yellowish throat and belly and slightly buffy vent.

106.3. Himalayan Bulbul
Pycnonotus leucogenys

20cm   p.338
Curly brown crest, black face and throat with white cheek and yellow vent; note white tail-tips.

106.9. Nicobar Bulbul
Hypsipetes nicobariensis

20cm   p.344
Dingy olive above with longish bill, blackish crown, whitish throat and yellowish belly.

106.4. White-eared Bulbul
Pycnonotus leucotis

20cm   p.337
Paler and less crested than Himalayan, with larger white cheek-patch and blacker crown.

106.10. Mountain Bulbul
Hypsipetes mcclellandi

23cm   p.342
Large; olive with shaggily crested chestnut head and buffier belly, with usually ruffled greyish throat.

106.5. Himalayan Black Bulbul
Hypsipetes leucocephalus

23cm   p.342
Large and slaty with red bill and legs, ragged crest, black crown and moustache, and notched tail.

106.11. White-throated Bulbul
Alophoixus flaveolus

23cm   p.341
Chunky and stout-billed, with olive mantle and rufous crown, wings and tail, spindly crest, whitish face and throat, and yellow belly.

106.6. Square-tailed Black Bulbul
Hypsipetes ganeesa

23cm   p.344
Like Himalayan Black but head less crested, with no black moustache; tail squared. Larger, bigger-billed, and paler in Sri Lanka than in S India.

PLATE 107. YELLOW BULBULS (J. Anderton)

Bulbuls Pycnonotidae (continued from Plate 106)

107.1. **Black-crested Bulbul**
Pycnonotus flaviventris

18cm p.336
Black head with long slender crest and pale yellow eyes.

107.2. **Black-capped Bulbul**
Pycnonotus melanicterus

17cm p.337
Uncrested black cap, dark eye, yellow throat and white tail-tips.

107.3. **Flame-throated Bulbul**
Pycnonotus gularis

18cm p.336
Smaller than Black-crested, with a short ragged crest, orange-red throat, and small whitish tail-corners.

107.4. **Black-headed Bulbul**
Pycnonotus atriceps

18cm p.336
Small; olive with uncrested black head, blue eye, and yellow belly, rump and tail-tip; note black primaries and subterminal tail-band. Rare grey morph lacks some or all yellow and green.

107.5. **Andaman Bulbul**
Pycnonotus fuscoflavescens

18cm p.336
Similar to allopatric Black-headed but duskier with ill-defined blackish-olive face.

107.6. **Yellow-throated Bulbul**
Pycnonotus xantholaemus

20cm p.338
Silvery-grey with yellow-olive head, wings and tail, and yellow throat and vent.

107.7. **Grey-headed Bulbul**
Pycnonotus priocephalus

19cm p.335
Olive with grey head and hindquarters, yellow forecrown, whitish eye and pale green bill. Tail has dark outer feathers.

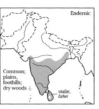

107.8. **White-browed Bulbul**
Pycnonotus luteolus

20cm p.341
Drab olive and unkempt, with white brow and cheek-streaks, yellowish throat and vent.

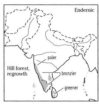

107.9. **Yellow-browed Bulbul**
Iole indica

20cm p.341
Plain bright olive with black bill, and yellow brow, face and underparts.

107.10. **Yellow-eared Bulbul**
Pycnonotus penicillatus

20cm p.338
Black head with ornate white lore-plumes, yellow ear-tuft and cheek-spot, and white throat. Otherwise plain olive with yellow underparts.

107.11. **Crested Finchbill**
Spizixos canifrons

20cm p.335
Olive, with pale conical bill, and grey head with blackish mask and crest, yellowish belly, and black terminal tail-band. Juvenile is drabber and weakly crested.

107.12. **Flavescent Bulbul**
Pycnonotus flavescens

20cm p.340
Large and olive with heavy tail; head grey with short occipital crest, whitish spectacles and throat, and yellowish belly and vent.

107.13. **Striated Bulbul**
Pycnonotus striatus

20cm p.335
Large and boldly streaked white below, with long pointed crest, and yellow eye-ring, throat and vent.

PLATE 108. IORAS, LEAFBIRDS AND FAIRY-BLUEBIRD (J. Anderton)

Ioras *Aegithina*

Rather dumpy insectivores with longish, silvery bills and pale eyes. Superficially like much larger orioles. Note wing and tail pattern.

108.1. Common Iora
Aegithina tiphia

14cm   p.344
Bright yellow below with large white wing-bar and short square dark tail; tertial trim if present fine and yellowish. Breeding male has black wings and tail; entirely black above in S, green above with black cap in N. Female is olive above; non-breeding male similar but with blacker wings and tail.

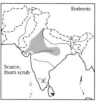

108.2. Marshall's Iora
Aegithina nigrolutea

14cm   p.346
From Common by broad white edges and tips on tail feathers and tertials. Breeding male is pale olive above with yellowish collar and black cap. Female and non-breeding male are pale olive with pale yellow head.

Leafbirds *Chloropsis*

Mainly green arboreal birds with slender, slightly decurved bills and square tails. Note extent of black bib and blue on throat, and colour of head.

108.3. Orange-bellied Leafbird
Chloropsis hardwickii

19cm   p.348
Diagnostic orange-ochre belly in both sexes. Male has dark blue wings and tail; female has blue moustache and green throat. Juv. green with silvery-blue moustache, and usually with some ochre feathers below.

108.4. Gold-fronted Leafbird
Chloropsis aurifrons

19cm   p.346
Bright orange forehead, with most of cheek black; sexes similar. In N, throat blue; in S, moustache only blue. Juv. has yellowish forecrown, blue moustache and no mask.

108.5. Jerdon's Leafbird
Chloropsis jerdoni

19cm   p.348
No orange on forehead; like allopatric Blue-winged but duller, with dirty yellow collar on cheek and neck, and at most a little blue in wing and tail. Male's black mask more limited and blue moustache bigger than in Gold-fronted; female has well-defined turquoise throat-patch.

108.6. Blue-winged Leafbird
Chloropsis cochinchinensis

18cm   p.346
Similar to Jerdon's, but with blue flight feathers and tail. Male has lemon-yellow collar enclosing mask; female has diffuse blue-green throat and golden-tinged nape.

Fairy-bluebird *Irena*

Rather large and heavily built; can recall *Aplonis* starlings. Formerly thought related to leafbirds, but systematic position now uncertain.

108.7. Asian Fairy-bluebird
Irena puella

27cm   p.589
Dark and stocky, with red eye and heavy black bill. Male black with blue crown, mantle, rump and vent; female (and imm.) dusky slate-blue with blacker wings and tail.

PLATE 109. SHRIKES (J. Anderton)

Shrikes *Lanius*

Predatory, hook-billed songbirds, usually masked. Immatures have obscured, barred plumages.

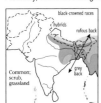

109.1. 'Black-headed' Long-tailed Shrike
Lanius schach tricolor group

25cm   p.350
Large with black cap and long black tail, rufous mantle, flanks and rump, and small white carpal patch. Imm. grey and barred above, lightly scaled below.

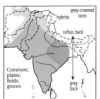

109.2. 'Rufous-backed' Long-tailed Shrike
Lanius schach erythronotus group

Like 'Black-headed' but crown grey. W form also has rufous on rear mantle; Gujarat and S forms are paler-crowned with only rump rufous. Imm. grey and barred above, lightly scaled below.

109.3. Southern Grey Shrike
Lanius meridionalis

25cm   p.352
Large, pale, with limited white in wing and grey rump. NW *pallidirostris* paler, duller, with white brow. Imm. brownish, finely barred below.

109.4. Great Grey Shrike
Lanius excubitor   25cm   p.352

Vagrant; NW (not illustrated). Like *pallidirostris* Southern Grey but greyer, rump white, much white in wing.

109.5. Grey-backed Shrike
Lanius tephronotus

25cm   p.351
Dark grey above with brown tail, and rufous on rump and flanks but not mantle; no white carpal patch. Imm. with heavy blackish bars above and scales below. In NW, hybridises with 'Rufous-backed'.

109.6. Lesser Grey Shrike
Lanius minor

23cm   p.351
Stout, with heavy black bill; large carpal patch (but no white trim) and pinkish underparts. Broad black forecrown in male. Imm. browner with pale forehead and pale-based bill.

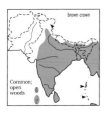

109.7. Brown Shrike
Lanius c. cristatus

19cm   p.349
Brown with rufous tail and flanks, and black mask below pale brow; no carpal patch. In E, vagrant *superciliosus* bright chestnut above with strong white forehead.

109.8. 'Philippine' Shrike
Lanius cristatus lucionensis

19cm   p.349
Greyer above than Brown, with pale grey crown.

109.9. Rufous Shrike
Lanius phoenicuroides

17cm   p.349
Smaller, brighter than Brown, with white carpal, whiter underparts.

109.10. Isabelline Shrike
Lanius isabellinus

17cm   p.349
Pale sandy-brown with rufous tail, diffuse mask and minimal carpal patch.

109.11. Bay-backed Shrike
Lanius vittatus

17cm   p.350
Small and slim, with pale grey rump. Male has black forecrown, whitish nape, chestnut mantle, and large carpal patch on black wing. Female much duller and paler.

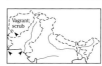

109.12. Woodchat Shrike
Lanius senator

17cm   p.352
Black above with rusty nape, white scapulars, carpal patch and rump. Juv. pale grey-brown above scaled blackish, with darker cheek-patch and whitish supercilium, scapulars and rump.

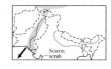

109.13. Red-backed Shrike
Lanius collurio

17cm   p.349
Small, with short squared tail; no carpal patch. Female much duller and scalier than distinctive male.

109.14. Burmese Shrike
Lanius collurioides

22cm   p.350
Like Bay-backed but darker, with chestnut rump. Female duller, paler than male.

PLATE 110. WAXWING, HYPOCOLIUS, MONARCHS, FANTAILS AND DIPPERS (J. Anderton)

Waxwing and Hypocolius *Bombycilla, Hypocolius*

Sleek, mid-sized passerines with short bills and broad gapes. Sexes similar in *Bombycilla*, moderately different in *Hypocolius*; juveniles duller in both.

110.1. Bohemian Waxwing
Bombycilla garrulus

18cm   p.332
Tawny-grey and crested, with black mask and throat, yellow and white trim on wings, rufous vent and broad yellow tail-tip.

110.2. Grey Hypocolius
Hypocolius ampelinus

25cm   p.332
Slim and pale brownish-grey with long, black-tipped tail. Male has blackish mask and broad white wing-tip.

Monarchs *Hypothymis, Terpsiphone*

Diverse, attractive mid-sized passerines, mostly with metallic-looking, rather broad flycatcher-like bills and elaborate plumage; in region sexes usually differ strongly, juveniles like females.

110.3. Black-naped Blue Monarch
Hypothymis azurea

16cm   p.333
Flycatcher-like; cobalt-blue head with black bill and nape-patch. Male is blue with narrow black collar; female duller, brownish above and greyish on flanks.

110.4. Asian Paradise Flycatcher
Terpsiphone paradisi (part)

N races 20cm/ 50cm w/streamers   p.332
Slim with glossy black crown and crest, lead-blue eye-ring and bill. Male has long tail-streamers and entirely black head; either wholly white, or bright rufous above with greyish breast. Female and imm. like latter, but with grey throat and no streamers. Crest length varies by region; see text.

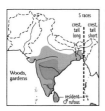

110.5. *Terpsiphone paradisi* (part)
S races 20cm/ 50cm w/streamers   p.332

Fantails *Chelidorhynx, Rhipidura*

Flycatcher-like birds with very broadly rounded and white-tipped tails, spread demonstratively. Note underpart colour and pattern.

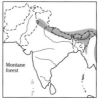

110.6. Yellow-bellied Fantail
Chelidorhynx hypoxantha

8cm   p.335
Tiny, olive with dark mask, yellow brow and underparts; tail has white trim and small tips. See Black-faced Warbler (Pl. 149).

110.7. White-browed Fantail
Rhipidura aureola

18cm   p.334
Long tail with very large white tips; broad white brow, wing-bar; white below with blackish gorget.

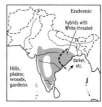

110.8. White-spotted Fantail
Rhipidura albogularis

17cm   p.334
From White-throated by longer white brow, whitish-buff belly, white-spotted breast-band and very reduced pale tail-tips.

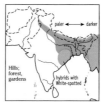

110.9. White-throated Fantail
Rhipidura albicollis

17cm   p.334
Dark belly and vent, short white brow and large white tail-tips.

Dippers *Cinclus*

Chunky and short-tailed birds, usually near running water.

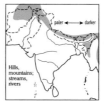

110.10. Brown Dipper
Cinclus pallasii

22cm   p.353
Large and entirely reddish-brown; juv. paler and greyer, scaled overall.

110.11. White-throated Dipper
Cinclus cinclus

20cm   p.353
Smaller than Brown, grey with white bib; juv. often whitish below and weakly scaled, with pale spectacles. Rare dark morph ('*sordidus*') is entirely grey.

PLATE 111. COCHOAS, GRANDALA, ROCK- AND WHISTLING-THRUSHES (T. Schultz)

Cochoas *Cochoa*

Stout, short-billed and -legged, lavishly coloured forest species.

111.1. Purple Cochoa
Cochoa purpurea

28cm   p.371

Blackish mask and silver-blue crown, tail and wing-patches. Male purplish-grey; female brown.

111.2. Green Cochoa
Cochoa viridis

28cm   p.372

Moss-green with blue crown and tail, and silvery-and-black wings (stained rufous in female).

Grandala *Grandala*

Alpine bird, rather like rock-thrushes but slimmer and longer-winged; more aerially adapted.

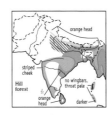

111.3. Grandala
Grandala coelicolor

23cm   p.389

Small, long-winged and fine-billed. Male purplish-blue; female grey-brown with pale streaks.

Rock-thrushes *Monticola*

Stout, grey-and-rufous thrushes of open habitats.

111.4. Chestnut-bellied Rock-thrush
Monticola rufiventris

24cm   p.388

Large. Male dark blue with blackish mask, chestnut belly. Female greyish-brown with buff lores, throat and border to dark cheek, and blackish scales on buffy underparts; from female *Zoothera* by dark legs, dark vent.

111.5. Rufous-tailed Rock-thrush
Monticola saxatilis

19cm   p.388

Short rufous tail. Male has blue head, white rump, rufous underparts; female is buff barred dark above and below.

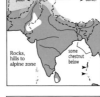

111.6. Blue Rock-thrush
Monticola solitarius

23cm   p.389

Dull slate-blue. Male has white spotty wing-bar; female finely scaled grey-and-buff below.

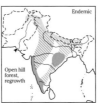

111.7. Blue-headed Rock-thrush
Monticola cinclorhynchus

17cm   p.388

Small. Male has blue head and shoulder, black mask, and orange-rufous underparts and rump, and white wing-patch; female greyish-brown with weak face pattern, dark scales below. In extreme SE, see text for possible White-throated *M. gularis* (p.388).

Ground-thrush *Zoothera* (others on Plate 114)

111.8. Orange-headed Thrush
Zoothera citrina

21cm   p.356

Small; grey with rufous head and underparts. Female has olive tint above. Mainland birds have white bar on wing; in Peninsula, white face and dark cheek-bars. Andamans and Nicobars races have pale throats, no wing-bars.

Whistling-thrushes *Myophonus*

Dark thrushes of riverine habitats; tremendous songsters. No overlap between ranges.

111.9. Ceylon Whistling-thrush
Myophonus blighi

20cm   p.370

Small and short-tailed with blue shoulder. Male blackish-blue; female dark brown.

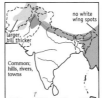

111.10. Blue Whistling-thrush
Myophonus caeruleus

33cm   p.371

Very large with yellow bill; deep blue spangled with silver.

111.11. Malabar Whistling-thrush
Myophonus horsfieldii

25cm   p.371

Large with black bill; deep blue with blackish foreparts, blue forehead and shoulder.

PLATE 112. *TURDUS* THRUSHES I (T. Schultz)

Thrushes *Turdus* (continued on Plate 113)

A large assemblage of 'typical' thrushes, generally more simply patterned and less colourful than other genera. Females are similar to but duller than males. Note face pattern, colour of bill and legs.

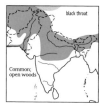

112.1. Black-throated Thrush
Turdus [ruficollis] atrogularis
25cm   p.366
Warm grey above with whitish underparts, no rufous in tail. Male has black face and breast; female has whitish brow, face and throat with mottled malar and breast.

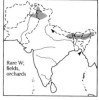

112.6. Dusky Thrush
Turdus [naumanni] eunomus
24cm   p.366
White brow and throat with dark mask, checkered breast and flanks, and chestnut wings.

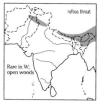

112.2. Red-throated Thrush
Turdus [r.] ruficollis
25cm   p.365
From Black-throated by rufous in outertail. Male has rufous brow, face and breast; female and first-year rufous-tinged breast.

112.7. Black-breasted Thrush
Turdus dissimilis
22cm   p.362
Yellow bill, eye-ring and legs, and rufous-orange lower breast and flanks, with white belly and vent. Male has black hood, dark brown in female with whitish throat and dark malar.

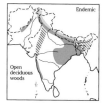

112.3. Tickell's Thrush
Turdus unicolor
21cm   p.362
Male pale ashy-grey, female pale olive-brown above, clay-brown on flanks and spotted breast, with whitish throat and streaked malar; both have yellow bill and legs, white belly to vent.

112.8. Mistle Thrush
Turdus viscivorus
28cm   p.367
Large; recalls some *Zoothera* thrushes (Pl. 114), but greyer above with white tail-tips, and more rounded spots below.

112.4. Grey-sided Thrush
Turdus feae
23cm   p.365
Recalls Eyebrowed but head and upperparts russet brown. Male has silvery-grey breast and flanks, fawn on female.

112.9. Song Thrush
Turdus philomelos
24cm   p.367
Warmer brown above than much larger Mistle, with no white tips to tail.

112.5. Eyebrowed Thrush
Turdus obscurus
23cm   p.365
Olive brown above with white brow, chinstrap and belly. Male has slaty-grey hood and fulvous breast and flanks; female paler and browner-headed with whitish throat, dark malar.

112.10. Fieldfare
Turdus pilaris   27cm   p.366
Hypothetical; Uttar. Bluish-grey head and rump, chestnut mantle; buffy-white and heavily spotted below.

112.11. Redwing
Turdus iliacus
22cm   p.367
Relatively small; brown with long white supercilium, heavily streaked throat and breast, and bright rufous flanks and wing-lining, bicoloured bill.

PLATE 113. *TURDUS* THRUSHES II AND BLACKBIRDS (T. Schultz)

Thrushes and **blackbirds** *Turdus* (continued from Plate 112)

Most species here are particularly simply coloured and patterned, without spots, wing-bars, supercilia, etc.

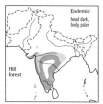

113.1. Indian Blackbird
Turdus simillimus
nigropileus group

<26cm   p.364

Mid-sized and grey with yellow-orange bill, eye-ring and postorbital patch, and legs. N Peninsular male is smoky with blackish cap; female duller and browner.

113.6. Ring Ouzel
Turdus torquatus

p.362

Possible; extreme NW (not illustrated). Like Common Blackbird but with white to pale brown chest crescent, and broad white wing edgings. Male black, other plumages browner.

113.2. *Turdus simillimus* S races

Male almost blackish in S Peninsula; in Sri Lanka, both sexes blue-grey and rather scaly in appearance.

113.7. Grey-winged Blackbird
Turdus boulboul

28cm   p.363

Dark with yellow bill and legs. Male is black with large pale grey wing-panel, female olive-brown with a diffuse rufescent wing-panel.

113.3. Common Blackbird
Turdus merula

26cm   p.363

Smaller than Tibetan. Male has yellow bill and eye-ring; female is slaty-brown with diffuse pale lores and eye-ring, dusky bill, and paler below with dark streaks on whitish throat.

113.8. Chestnut Thrush
Turdus r. rubrocanus

27cm   p.364

Chestnut with pale grey head, yellow bill and legs, and dark wings and tail.

113.4. Tibetan Blackbird
Turdus maximus

>26cm   p.363

Very large and dark with black legs and no eye-ring. Male is black with yellow bill; female blackish-brown with duskier bill.

113.9. *Turdus rubrocanus gouldi*

Dark-headed form hypothetical for NE.

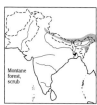

113.5. White-collared Blackbird
Turdus albocinctus

27cm   p.362

Dark with yellow bill and legs; very broad collar is white in male, greyish in female.

113.10. Kessler's Thrush
Turdus kessleri

27cm   p.345

Large, with dark hood, wings and tail. Male has black hood, buff mantle and breast with dark chestnut rear-parts; female is duller and duskier, with same diagnostic pattern.

PLATE 114. *ZOOTHERA* THRUSHES (T. Schultz)

Thrushes *Zoothera* (see also Plate 111)

Usually shorter-tailed, dumpier and larger-billed than *Turdus* thrushes, coloured in browns and blacks, with a prominent whitish wing-stripe showing in flight; also more terrestrial in behaviour. Note pattern of face, wings and vent, details of bill.

114.1. Spot-winged Ground-thrush
Zoothera spiloptera

21cm p.357
Small; double wing-bar, black vertical patches on white face, and rounded spots below.

114.2. Nilgiri Thrush
Zoothera neilgherriensis

26cm p.359
Smaller, bigger-billed and shorter-tailed than Small-billed, with two buffy wing-bars and no strong pale spangles on darker russet upperparts.

114.3. White's Thrush
Zoothera aurea >26cm p.358

Hypothetical; NE. From Small-billed by longer bill and bolder, more rounded scales above, stronger eye-ring and malar spots, weaker blackish ear-patch and stronger, whiter wing-bar.

114.4. Plain-backed Thrush
Zoothera mollissima

27cm p.357
Very like Long-tailed, but face and wings plainer; more crescentic spots below reach vent.

114.5. Long-tailed Thrush
Zoothera dixoni

27cm p.358
From very similar Plain-backed by dark ear-patch bordered by buff, pale double wing-bars, and more rounded spots below with unmarked belly and vent.

114.6. Ceylon Scaly Thrush
Zoothera imbricata

<26cm p.359
Shorter-tailed and bigger-billed than Nilgiri, washed cinnamon below with unscaled belly and vent.

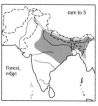

114.7. Small-billed Scaly Thrush
Zoothera dauma

26cm p.358
Black scales and pale spots on upperparts, heavy scales below with white belly and vent. Bill relatively small.

114.8. Pied Ground-thrush
Zoothera wardii

22cm p.356
Male vividly pied with yellow bill and legs; female from female Siberian by dark cheek not surrounded by buff and less buffy tone overall.

114.9. Siberian Thrush
Zoothera sibirica

22cm p.357
Dark with scaled pale vent. Male is slate with white brow, belly and tail-tips and dark bill; female dark brown with long buff brow bordering cheek, and dark scales on buff underparts.

114.10. Long-billed Ground-thrush
Zoothera monticola

28cm p.359
Very stout, short-tailed and dark with massive dark bill, almost entirely slaty-brown with dusky ochre throat.

114.11. Dark-sided Ground-thrush
Zoothera marginata

25cm p.359
From Long-billed by smaller bill, blackish ear-patch with pale rear border, and strongly scaled flanks.

PLATE 115. SHORTWINGS AND BLUE ROBINS I (J. Anderton)

Shortwings *Brachypteryx, Heteroxenicus*

Long-legged and very short-tailed terrestrial chats, usually very elusive. Compare the browner female plumages with small forest babblers, especially *Pellorneum* and *Malacocincla* (Pl. 139).

Montane forest, thickets

115.1. White-browed Shortwing
Brachypteryx montana

13cm   p.390
Male blackish-blue with bold white brow; female dark olive-brown above with chestnut eye-ring and forehead, and greyer below with rufous vent.

Near-endemic
Scarce; tall grass, bamboo

115.3. Rusty-bellied Shortwing
Brachypteryx hyperythra

13cm   p.389
Dark above and orange-rufous below with dark bill and long pink legs. Male dusky blue above with short white brow (often hidden), female slaty-brown above.

males blue
Montane forest, regrowth
males brown

115.2. Lesser Shortwing
Brachypteryx leucophrys

13cm   p.390
Small with long pink legs and white brow (often hidden). Male in Himalayas is blue-grey above with greyish sides; in S Assam hills, brown above with some scales on flanks and often on breast. Female is dark brown above, more russet on rear, with breast-band and flanks scaled fulvous-brown. See Spot-throated Babbler (Pl. 139), which has buffier belly and vent.

Scarce; alpine screes; winters rocky gullies

115.4. Gould's Shortwing
Heteroxenicus stellatus

13cm   p.389
Reddish-chestnut above and grey with black vermiculations and white stars below.

Blue robins *Cinclidium, Myiomela*

Rather large, long-legged chats found on or near the ground in forest undergrowth and thickets. *Callene* has been treated previously as a single species placed in *Brachypteryx*. Note tail pattern, colour of belly.

Scarce; bamboo in heavy forest

115.5. Blue-fronted Blue Robin
Cinclidium frontale

19cm   p.400
Male is slate-blue with pale brow over black lores, and blue shoulder; darker in NE, with reduced greyish scales on belly and vent. Female is dark brown above, rich fulvous below with whitish throat and centre of belly and white-tipped vent. Female Hodgson's Blue Robin included for comparison; see Pl. 120.

Endemic
Montane shola forest, thickets

115.7. White-bellied Blue Robin
Myiomela albiventris

15cm   p.400
Chunky and dark; slate-blue with whitish brow over black lores, blue-grey flanks and clear-cut white belly and vent. Male of sympatric White-bellied Blue Flycatcher (Pl. 126) is brighter blue, much shorter-legged, and lacks white brow.

Common; hill forest; winters lower

115.6. White-tailed Blue Robin
Myiomela leucura

17cm   p.400
Broad white basal panels on dark tail. Male is black with white neck-spot and blue forehead and shoulder; female olive-brown with buffy eye-ring, rufescent breast and whitish throat.

Endemic
Montane shola forest, thickets

115.8. Nilgiri Blue Robin
Myiomela major

15cm   p.400
Slightly paler and bluer than White-bellied, with bluish brow over dusky lores, fulvous flanks and vent, and buffy-white mid-belly.

Plate sponsored by Mr. Wallace Dayton.

PLATE 116. BUSH- AND BLUE ROBINS II (J. Anderton)

Bush-robins *Tarsiger, Luscinia*

Small terrestrial chats, some of which resemble certain rufous-breasted flycatchers (Pl. 124–125), but are plumper and distinctly longer-legged. Note colour of throat and belly, presence of supercilium; in females, colour of tail.

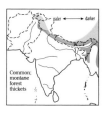

116.1. Himalayan Red-flanked Bush-robin
Tarsiger rufilatus

15cm   p.393

Note orange flank-patch, narrow white throat-streak and blue tail in both sexes. Male is blue above and on side of breast, with brighter brow and shoulder; female is brown above with pale eye-ring.

116.2. Northern Red-flanked Bush-robin
Tarsiger cyanurus   15cm   p.394

Vagrant; NE. Shorter-tailed and buffier below than Himalayan Red-flanked, male with whiter brow. Female much warmer brown. See text.

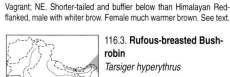

116.3. Rufous-breasted Bush-robin
Tarsiger hyperythrus

13cm   p.394

Tiny bill, white vent and blue tail. Male deep blue above with orange throat-stripe, breast and belly; female dark brown above and on face with buff eye-ring and throat-stripe.

116.4. White-browed Bush-robin
Tarsiger indicus

15cm   p.394

Long-tailed. Male is slate (not blue) above with long white brow, and rusty below; female is fulvous-brown including tail, with white brow often reduced or lacking, and buffy-ochre underparts.

116.5. Siberian Blue Robin
Luscinia cyane

15cm   p.393

Short-tailed with long pale legs. Male blue above with blackish mask and side-stripe, and white below; female dull brown with buffy eye-ring, whitish below with scaly breast-band.

116.6. Indian Blue Robin
Luscinia brunnea

15cm   p.393

Stout, neckless and short-tailed, recalling shortwing. Male blue with white brow, blackish face, rufous-orange breast and flanks, and white vent; female olive-brown above with rufescent rump, whitish below with weak fulvous mottled breast and flanks.

116.7. Firethroat
Luscinia pectardens

15cm   p.392

Skulking, with longish bill, rounded tail, and long dark legs. Male like male Siberian Blue, with orange throat, white neck-spot and basal tail-panels; female is warm brown above with slightly rufous rump and tail, and buffy below.

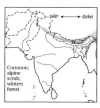

116.8. Golden Bush-robin
Tarsiger chrysaeus

15cm   p.394

Small and chunky. Male has vivid golden-orange rump, tail-base, scapulars, broad brow and underparts, with black mask and tail and dark mantle and dark wings. Female olive with yellowish brow and underparts, some orange on side of tail.

242

PLATE 117. BROWN ROBINS, BLUETHROAT AND RUBYTHROATS (J. Anderton)

Robin, Bluethroat, Rubythroats, Nightingale, Scrub-robin *Erithacus, Luscinia, Cercotrichas*

Diverse brown-backed chats, most with bright throat-patches.

117.1. European Robin
Erithacus rubecula

13cm   p.390
Small and rotund, with orange-rufous face and breast and white belly.

117.2. Bluethroat
Luscinia s. svecica

15cm   p.392
Smoky-grey, with white brow and belly, and rufous basal panels on blackish tail. Breeding male has blue breast, obscured by pale tips in fresh plumage; female has white throat with heavy blackish malar and blotchy necklace. See text for racial variation.

117.3. *Luscinia svecica pallidogularis*

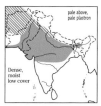

117.4. *Luscinia svecica abbotti*

117.5. Himalayan Rubythroat
Luscinia pectoralis (part)

15cm   p.391
Greyer (especially on breast) than Siberian, with white tail-tips. Male has white brow, broad black breast-band (obscured somewhat when fresh), and white basal panels and tips on black tail. Female is grey-brown with whitish brow, submoustachial and throat.

117.6. *Luscinia pectoralis tschebaiewi*

Male of race *tschebaiewi* has white submoustachial like Siberian.

117.7. Siberian Rubythroat
Luscinia calliope

15cm   p.391
Browner than Himalayan, with solidly dark tail. Male has diffusely grey rather than solid black breast, and white brow and submoustachial (but note race *tschebaiewi* of Himalayan). Female from female Himalayan by brownish breast-band.

117.8. Common Nightingale
Luscinia megarhynchos

18cm   p.391
Plain brown with rufescent rump and tail, and whitish below with buffy breast. Juv. has whitish wing-bars and tertial-trim.

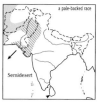

117.9. Rufous-tailed Scrub-robin
Cercotrichas galactotes

17cm   p.395
Sandy with paler brow, dark eyeline and whisker, rufous rump and longish rufous tail (commonly cocked) with black subterminal band and broad white tips.

PLATE 118. SHAMAS, BLACK ROBIN, ROCK-CHAT AND FORKTAILS (J. Anderton)

Shamas and magpie-robin Copsychus

Essentially pied, long-tailed chats with melodious voices.

118.1. Andaman Shama
Copsychus albiventris

23cm   p.396
Like White-rumped but shorter-tailed, and white below with rufous flanks; sexes similar.

118.2. White-rumped Shama
Copsychus malabaricus

25cm   p.395
Long-tailed, with rufous belly, white rump and white outertail feathers. Male black above, female grey.

118.3. Oriental Magpie-robin
Copsychus saularis

20cm   p.395
Long-tailed, with white belly, wing-slash and tail-sides; male glossy black above, female grey.

Robin and Chat Saxicoloides, Cercomela

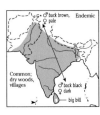

118.4. Indian Black Robin
Saxicoloides fulicatus

16cm   p.396
Dark and lanky, with chestnut vent. Male glossy black with white shoulder-flash; in NW, dark brown above. Female brown with blackish tail; darker in Sri Lanka.

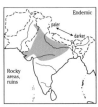

118.5. Brown Rock-chat
Cercomela fusca

17cm   p.408
Similar to female Indian Black Robin, with shorter, darker tail, no chestnut on vent.

Forktails Enicurus

Showy, pied, non-dimorphic chats always found near running water. Four species have long, ornate, forked tails. Calls are high and shrill, like those of whistling-thrushes. Note pattern of head (spectacles vs forecrown-patch) and colour of mantle.

118.6. White-crowned Forktail
Enicurus leschenaulti

28cm   p.401
Large, with white forecrown, black breast and solid black mantle. Juv. (not shown) like juv. Spotted.

118.7. Little Forktail
Enicurus scouleri

12cm   p.401
Very small with short, notched, white-sided black tail, and white rounded forecrest.

118.8. Slaty-backed Forktail
Enicurus schistaceus

25cm   p.401
White spectacles and slate-grey crown and mantle. Juv. lacks white forecrown, and has white eye-ring and blurry grey streaks on breast.

118.9. Black-backed Forktail
Enicurus immaculatus

25cm   p.401
From nearly identical Slaty-backed by black crown and mantle. Juvenile (not shown) is like juv. Slaty-backed but darker and browner above and more scaled below.

118.10. Spotted Forktail
Enicurus maculatus

26cm   p.402
Very like White-crowned, but with white-spangled hindcollar and mantle. Juv. entirely dark brown with weak pale streaks on throat and breast.

PLATE 119. TYPICAL REDSTARTS (L. McQueen)

Redstarts *Phoenicurus* (continued on Plate 120)

Smallish insectivorous chats recalling flycatchers in posture and behaviour; most species flicker or flirt their tails. Winter males have buffy feather-edges obfuscating field marks. Note wing pattern, and back, crown and belly colour; in females, eye-ring and throat colour are especially important.

119.1. Black Redstart
Phoenicurus ochruros rufiventris

15cm p.397

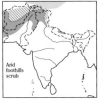

119.5. Eversmann's Redstart
Phoenicurus erythronotus

15cm p.397
Male has white wing-slash and diagnostic rufous throat and mantle. Female is plain grey-brown with buffy eye-ring, whitish tertial trim and double wing-bar.

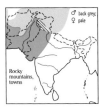

119.2. *Phoenicurus ochruros phoenicuroides*

Male is dark above and on breast with dark rufous rump and underparts; NW breeding race *phoenicuroides* has grey crown and lower mantle, black in C and NE wintering race *rufiventris*. Female *phoenicuroides* is drab brown with buffy belly; female *rufiventris* is much darker, with prominent brown breast-band and rufescent belly to vent. See female Blue-fronted (Pl. 120) which has different tail pattern.

119.6. Daurian Redstart
Phoenicurus auroreus

15cm p.398
Broad white wing-patch in both sexes. Male has frosty crown and black mantle. Female dark overall.

119.7. Hodgson's Redstart
Phoenicurus hodgsoni

15cm p.397
Male from similar male nominate Common by whitish wing-patch; upper breast black. Female is greyer than female Black and lacks strong rufous wash below, and is darker below than female Common.

119.3. Common Redstart
Phoenicurus p. phoenicurus

15cm p.397
Plain-winged. Male has white forecrown and lacks white in wing of Hodgson's. Female is sandier than female Black (both races), with whitish throat and belly.

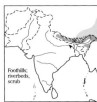

119.4. *Phoenicurus p. samamisicus*

Likely in extreme NW; as nominate but with a small white slash in wing; female paler and greyer above.

119.8. Blue-capped Redstart
Phoenicurus coeruleocephala

15cm p.399
Tail dark. Male is black with blue-grey cap, and white wing-slash and belly; female drab with dark chestnut rump and trim on outertail, and pale eye-ring, wing-bar and tertial trim.

PLATE 120. LARGE REDSTARTS, WATER CHATS, *IRANIA* AND *HODGSONIUS* (L. McQueen)

Redstarts *Phoenicurus* (continued from Plate 119)

120.1. Blue-fronted Redstart
Phoenicurus frontalis

15cm   p.398
Broad blackish T on rufous tail. Male has dark blue head, breast and mantle, and orange belly. Female has belly and vent washed yellow-orange; from female Black (Pl. 119) by tail pattern. See bush-robins (Pl. 116).

120.2. White-throated Redstart
Phoenicurus schisticeps

15cm   p.398
White throat-spot and wing-slash, and chestnut rump and basal tail-panels. Male has slaty-blue crown and chestnut underparts with white belly; female has buffy eye-ring.

120.3. White-winged Redstart
Phoenicurus erythrogastrus

18cm   p.398
Large with plain rufous tail. Male has white crown and wing-patch; female plain sandy-brown.

Redstart-like chats *Rhyacornis, Chaimarrornis*

120.4. Plumbeous Water-redstart
Rhyacornis fuliginosa

12cm   p.399
Small, squat, short-tailed and largely blue-grey. Male has rufous rump and tail; female is scaly grey-and-white below with white rump and basal tail-panels.

120.5. White-capped River-chat
Chaimarrornis leucocephalus

19cm   p.399
Black above with white crown, chestnut rump, tail and underparts; from White-winged Redstart by all-dark wing, and black tail-tip.

Robins *Irania, Hodgsonius*

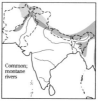

120.6. White-throated Robin
Irania gutturalis

18cm   p.396
Longish-billed and grey above with black tail and white throat, belly and vent. Male has white brow, black mask and bright rufous or buff breast and flanks. Female paler and duller with orange wash to breast and flanks.

120.7. Hodgson's Blue Robin
Hodgsonius phaenicuroides

19cm   p.399
Slim with longish rounded tail. Male slate-blue with white belly and rufous basal tail-panels (but not rump). Female olive-brown above with buffy eye-ring and dull rufous tail, buffy below with whitish throat and belly; less fulvous below than female Blue-fronted Blue Robin (Pl. 115).

PLATE 121. WHEATEARS (L. McQueen)

Wheatears *Oenanthe*

A confusing group of open-country chats, essentially terrestrial, in which males are usually boldly black and white, and females and subadults grey and/or sandy and white; fresh birds may look markedly different from worn ones. Note above all tail pattern, which is usually diagnostic in all plumages; note also upperpart patterns, especially in males, and facial pattern.

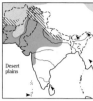

121.1. Isabelline Wheatear
Oenanthe isabellina

16cm p.408

Lankier, larger-billed than Northern, with narrow white rump-patch and no pronounced T on tail; pale and very pale-cheeked with brow wider before eye.

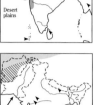

121.2. Northern Wheatear
Oenanthe oenanthe

15cm p.406

Broad black T on tail. Male blue-grey above, peach below; female browner, from female Isabelline by stronger brow, whiter behind eye, narrower black terminal tail-band.

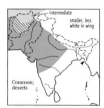

121.3. Desert Wheatear
Oenanthe d. deserti

15cm p.407

Only wheatear with only basal third of tail white, no T-pattern, and broad black terminal band.

121.4. *Oenanthe deserti oreophila*

More easterly-breeding race *oreophila* (not illustrated) is larger, with more white in wings.

121.5. Finsch's Wheatear
Oenanthe finschii

17cm p.406

Broad T on tail; rather large with slim bill. Male like male Desert, but paler above with black scapulars, and black of throat broadly connected with wing. Female as pale as Red-tailed, sandier above with wings mostly edged pale; throat dark or whitish.

121.6. Pied Wheatear
Oenanthe pleschanka

15cm p.406

Small and delicate with a fine bill; tail pattern is a rounded black W. Male has long white cap and white lower back. White-throated '*vittata*' morph rare. Female like female Desert but mantle darker.

121.7. Variable Wheatear
Oenanthe picata '*capistrata*'

16cm p.406

Larger-billed than Pied; evenly narrow terminal tail-band is straighter. Three forms, all with dark back and white rump. Male '*capistrata*' has white crown and nape not reaching upper mantle as in very similar Pied, and smaller rump-patch; female is drab brown with paler throat and breast grading to paler belly.

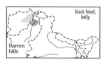

121.8. *Oenanthe picata* '*opistholeuca*'

Male '*opistholeuca*' is entirely black apart from rump, base of tail and vent (suggesting Pied Bushchat Pl. 122); female is dark brown in place of black.

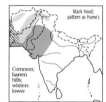

121.9. *Oenanthe p. picata*

Male *picata* has white breast and same pattern as larger, glossier Hume's; female is dark brown in place of black.

121.10. Mourning Wheatear
Oenanthe lugens

14cm p.407

Hypothetical; SW Pakistan. Similar to Pied but has with straight black terminal band on tail, and pale rufous vent. See text; sexes essentially similar.

121.11. Hooded Wheatear
Oenanthe monacha

17cm p.405

Long-billed and -bodied; largely white tail with black centre and corners. Male white-crowned with black back (beware Variable '*capistrata*'). Female pale with whitish brow and eye-ring, cinnamon-tinged cheek, rump and tail.

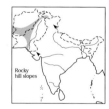

121.12. Hume's Wheatear
Oenanthe albonigra

18cm p.405

Large, heavy-billed, and mostly glossy black; like smaller, dumpier male Variable '*picata*', but white rump-patch extends to mid-back. Sexes alike.

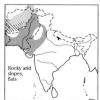

121.13. Red-tailed Wheatear
Oenanthe chrysopygia

16cm p.407

Drab with rufous rump, tail-base and tail-tip, with shallow dark T. Sexes similar.

PLATE 122. BUSHCHATS AND STONECHATS (J. Anderton)

Bushchats *Saxicola*

Tail-wagging, chunky chats favouring open areas and agriculture, usually on prominent, exposed perches. Males are usually boldly patterned in black and white (obscured by buffy or brown edges in winter), and females browner and streaked (and more difficult to identify). Note throat and rump colour, tail pattern, presence of brow.

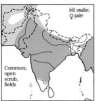

122.1. Pied Bushchat
Saxicola caprata (part)

13cm p.404
Male black with white rump, vent and (often hidden) scapulars; female drab with rufous rump and belly and blackish squared tail.

122.2. *Saxicola caprata nilgiriensis/atratus*

W Ghats *nilgiriensis* and Sri Lanka *atratus* larger; female darker and more heavily streaked; male not shown.

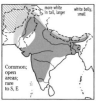

122.3. Common Stonechat
Saxicola torquatus indicus/ maurus

13cm p.403
Smallish with short square black tail, pale rump-patch, (often hidden) white badge in wing-coverts. Male when worn is black above and on throat, with rufous breast and (usually) white belly; in fresh plumage, upperparts look brown and streaked, throat pale, and underparts generally buffy. Female is streaked above and buffy-white below, without wing-bars; generally buffier than very greyish female White-tailed. Widespread *indicus* is small; male with white (worn) to pale buff belly.

122.4. *Saxicola torquatus stejnegeri*

In NE, *stejnegeri* is much like *indicus* but darker, with heavier bill.

122.5. *Saxicola torquatus przewalskii*

N winterer *przewalskii* is larger, with deep rufous-buff underparts.

122.6. White-tailed Stonechat
Saxicola leucurus

13cm p.403
Male has white tail-slots; when worn rufous breast-patch is smaller than in Common and often darker. Female is paler and greyer than female Common, with duller, smaller pale buff rump and dull brown tail with pale edges.

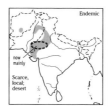

122.7. White-browed Bushchat
Saxicola macrorhynchus

15cm p.402
Slim dark bill; white on brow, scapulars, primary coverts and inner webs of tail feathers. Fresh male is buffy above with broad dark malar, white basal tail-panel; wears to blackish above. Female similar, without dark malar.

122.8. White-throated Bushchat
Saxicola insignis

17cm p.402
Large with long dark bill, pale rump. Male has white throat and wing-covert panel; female has dark streaks above, no sharp demarcation between throat and breast, and two buffy wing-bars. Juv. male like female, with white wing-covert panel.

122.9. Jerdon's Bushchat
Saxicola jerdoni

15cm p.405
Slim with longish tail, white below with no brow. Male black above; female dark brown above with russet rump.

122.10. Grey Bushchat
Saxicola ferreus

15cm p.405
Slim and longish-tailed, with broad white brow, dark mask and white throat. Worn male grey above with white wing-slash; female has bright rufous rump and tail. Fresh male paler, browner.

122.11. Whinchat
Saxicola rubetra 12cm p.404

Hypothetical; EC Afghanistan. Stocky, short-billed and short-tailed with buffy underparts, broad white brow below dark, streaked crown. Male has blackish mask.

PLATE 123. WHISTLER, BROWN AND PIED FLYCATCHERS (J. Anderton)

Whistler *Pachycephala*

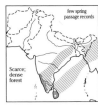

123.1. Mangrove Whistler
Pachycephala grisola

17cm   p.334

Chunky and large-headed with heavy black bill; pale grey with whitish throat and belly.

Jungle-flycatcher *Rhinomyias*

123.2. Nicobar Jungle-flycatcher
Rhinomyias nicobaricus

14cm   p.372

Longish, hooked bill with yellowish lower mandible, pinkish legs; dark brown above with buffy eye-ring, whitish below with mottled fulvous breast-band.

Flycatchers *Muscicapa, Ficedula*

Muscicapa are generally drab grey or brown and non-dimorphic flycatchers. *Ficedula* are small, chat-like, dimorphic species. Note especially facial pattern and bill colour.

123.3. Brown-breasted Flycatcher
Muscicapa muttui

13cm   p.375

Brown with rufescent wings, rump and tail, white spectacles, submoustachial and throat, and fulvous breast and flanks; note yellow lower mandible and legs.

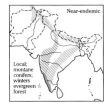

123.8. Rusty-tailed Flycatcher
Muscicapa ruficauda

14cm   p.373

Plain pale grey-brown with rufous rump and tail.

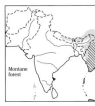

123.4. Dark-sided Flycatcher
Muscicapa sibirica

12cm   p.373

Small, short-billed and long-winged; dark olive-grey with bold white spectacles and throat.

123.9. Little Pied Flycatcher
Ficedula westermanni

10cm   p.378

Small, with dark bill and legs. Male pied with very bold white brow, wing-slash and basal tail-panels. Female very like female E Himalayan Ultramarine, but forehead less rufescent, breast-sides more mottled.

123.5. Ferruginous Flycatcher
Muscicapa ferruginea

10cm   p.375

Small; mostly rufous with slaty crown, pale spectacles, and white throat.

123.10 Semi-collared Flycatcher
Ficedula semitorquata   p.376

Possible, NW (not illustrated; map above). Male black above with white forehead, neck sides, wing patch, tail base, and underparts. Female pale, drab, with two white bars on brownish wing. See text for hypothetical Collared *F. albicollis* (p.375).

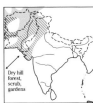

123.6. Spotted Flycatcher
Muscicapa striata

14cm   p.372

Rather large; very pale with dark bill, legs and eye, streaked crown and weak streaks on breast.

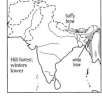

123.11. White-gorgeted Flycatcher
Ficedula monileger

11cm   p.377

Big-headed with short dark bill, pale legs; dusky olive-brown with white throat bordered black. Short brow is buffy in E Himalayas, white in S Assam hills.

123.7. Asian Brown Flycatcher
Muscicapa dauurica

14cm   p.373

Drab grey-brown with dark bill (lower mandible mostly pale), pale spectacles and underparts with greyish breast (looking variably streaked with wear).

123.12. Ultramarine Flycatcher
Ficedula superciliaris

10cm   p.378

Small, with dark bill and legs. Male dusky blue above and on side of breast, and white below; in W Himalayas also has white brow and basal tail-panels. Female is brownish-grey above with rufescent forehead; rump and tail rufous in E Himalayan female *aestigma*. Imm. male like female, with bluish mantle and tail, grey side of breast.

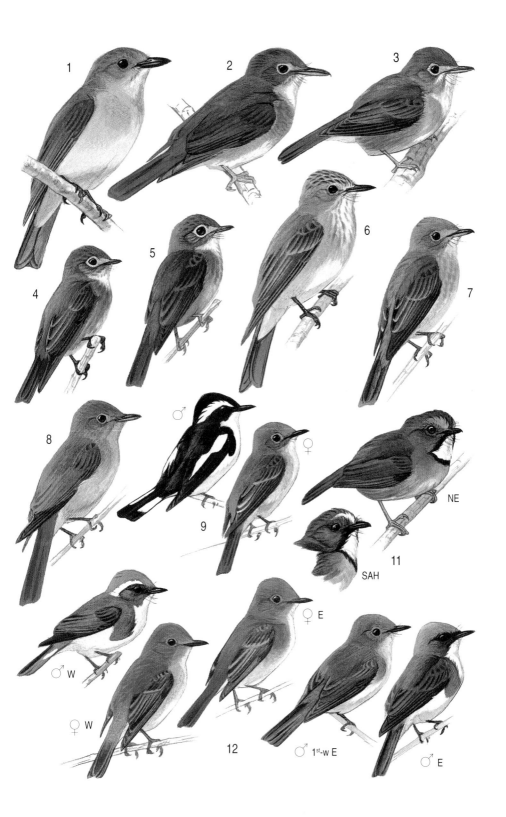

PLATE 124. GREY-BACKED FLYCATCHERS (J. Anderton)

Flycatchers Muscicapidae (continued from Plate 123, also Plates 125–126)

A large assemblage of mostly small insectivorous birds with broad flat bills and usually short tails; sit upright and sally from branches rather than glean foliage (as do most warblers). Some resemble longer-legged and more terrestrial chats such as bush-robins (Pl. 116). Generic relationships not yet understood, so they are presented here on basis of similarity in plumage. Note especially face and tail pattern, colour of rump and underparts.

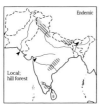

124.1. Kashmir Flycatcher
Ficedula subrubra

13cm　　　　　　　　　　　　　p.377
Darker, greyer above than Red-breasted. Male has long black malar framing chestnut throat; female whitish below with variably mottled rufous throat and breast.

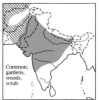

124.2. Red-breasted Flycatcher
Ficedula parva

13cm　　　　　　　　　　　　　p.377
Pale with white basal panels on blackish tail, whitish below with sandy breast and flanks; note pale lower mandible. Male has rufous-orange throat and greyish side of head.

124.3. Red-throated Flycatcher
Ficedula albicilla

13cm　　　　　　　　　　　　　p.377
From Red-breasted by blacker bill, tail and uppertail-coverts, and greyish breast. Male has paler rufous throat not reaching breast.

124.4. Black-and-orange Flycatcher
Ficedula nigrorufa

13cm　　　　　　　　　　　　　p.380
Rotund and short-tailed; orange-rufous with black cap and wings in male, dark brown in buff-spectacled female.

124.5. Orange-gorgeted Flycatcher
Ficedula strophiata

13cm　　　　　　　　　　　　　p.376
Black tail with white basal panels; olive-brown above with grey face and breast, fine white brow and rufous gorget (can be concealed); female may be much duller.

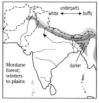

124.6. Slaty-blue Flycatcher
Ficedula tricolor

12cm　　　　　　　　　　　　　p.378
Prominent pale throat and dusky breast on white (Himalayas) or buff (S Assam hills) underparts. Male is bluish-slate above with white basal panels on black tail, black mask and frosty brow. Female olive-brown above with rufous rump and tail, below has white or buff throat and belly.

124.7. Slaty-backed Flycatcher
Ficedula hodgsonii

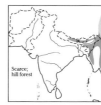

13cm　　　　　　　　　　　　　p.376
Small, long-tailed with tiny bill. Male dark slate above, and rufous below with paler vent. Female is very plain grey-brown, slightly paler on throat and belly, with slightly rufous tail.

124.8. Snowy-browed Flycatcher
Ficedula hyperythra

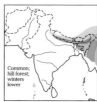

11cm　　　　　　　　　　　　　p.377
Small, short-tailed with russet secondaries. Male slaty above with white brow, rufous throat and breast. Female dusky olive-brown with dark-mottled face and breast, fulvous underparts.

124.9. Yellow-rumped Flycatcher
Ficedula zanthopygia

13cm　　　　　　　　　　　　　p.375
Vagrant; yellow rump and white wing-slash. Male black above with white brow, and yellow below. Female is olivaceous above with dilute, mottled yellowish underparts.

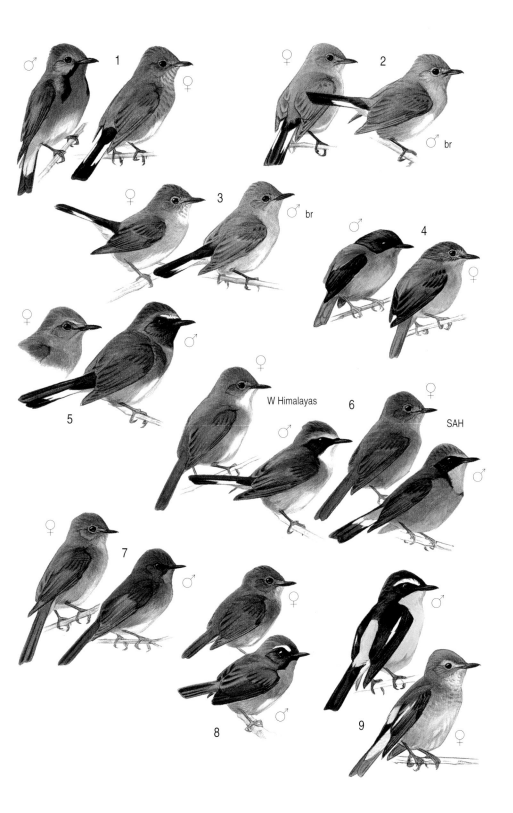

PLATE 125. BLUE-AND-RUFOUS FLYCATCHERS (J. Anderton)

Flycatchers Muscicapidae (continued from Plates 123–124, also Plate 126)

Most of these species are rufous below, with males largely bright blue above where females olive-brown (see also female White-bellied, Pl. 126); see Pl. 124 for similar slaty-backed rufous-breasted males. Note especially extent of rufous below and, in confusing females, colour and pattern of throat and cheek, and colour of tail.

125.1. Pygmy Blue Flycatcher
Muscicapella hodgsoni

8cm   p.387

Tiny with tiny bill, yellowish-rufous breast; short tail often cocked and wings drooped. Male blue above, brighter on brow; female pale olive-brown above with buffy eye-ring.

125.6. Large Blue Flycatcher
Cyornis magnirostris

15cm   p.385

Large with large bill, long wings. Male otherwise very like male Tickell's, but upperparts deeper blue; female like smaller female Blue-throated but side of head contrasts more with pale throat.

125.2. Sapphire Flycatcher
Ficedula sapphira

12cm   p.380

Small with tiny bill, orange-rufous throat and breast, and whitish belly. Breeding male deep blue above with bright blue forecrown; non-breeding has olive-brown head and mantle. Female olive-brown above with rufous rump and tail.

125.7. Hill Blue Flycatcher
Cyornis banyumas

14cm   p.385

Race *whitei*; see text. Like Tickell's in shape and bill size; male (not shown) is like male Large in colour. Female like female Large but tail less rufescent, bill smaller, wingtip shorter.

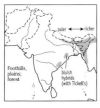

125.3. Pale-chinned Flycatcher
Cyornis poliogenys

14cm   p.384

Rather plain: olivaceous above with rufescent wings, rump and tail, grey head with whitish spectacles, and (rather dilute) rufous-ochre below with whitish throat and belly. Darker and brownish-crowned in NE; in E Ghats, see Tickell's for hybrids ('*vernayi*').

125.8. Vivid Niltava
Niltava [*vivida*] *oatesi*

18cm   p.383

Large, with no blue neck-spot. Male with dull blue patches above and peaked crown, dark rufous below with small indent into black throat. Female is dark above with buffy eye-ring and throat-streak, and grey-mottled below, variably washed rufous on breast. See female Large Niltava (Pl. 126).

125.4. Tickell's Blue Flycatcher
Cyornis tickelliae

14cm   p.385

Pale rufous-orange throat and diffuse flanks, whitish belly. Male blue above, bigger-billed in Sri Lanka. Female is ashy-blue above with bluer brow, wings and tail. Hybrids with Pale-chinned (*C. p.* '*vernayi*') in E Ghats are intermediate, greyer in NE Ghats, bluer in CE Ghats.

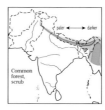

125.9. Rufous-bellied Niltava
Niltava sundara

15cm   p.383

Stout, with blue neck-spot. Male black above and on throat with blue crown, shoulder, rump and tail, orange-rufous below; female olive-brown with rufescent wings and tail, buffier face, and white upper breast-patch. See female White-tailed Blue Flycatcher (Pl. 126).

125.5. Blue-throated Flycatcher
Cyornis rubeculoides

14cm   p.384

In Himalayas, male like male Tickell's but throat blackish, rufous breast sharply marked from white belly. Female olive-brown above with buffy spectacles, slightly chestnut tail; lacks strong cheek/throat contrast of female Large. In S Assam hills, has rufous V on throat, more rufescent sides; see text

PLATE 126. BLUE FLYCATCHERS (J. Anderton)

Flycatchers Muscicapidae (continued from Plates 123–125)

Most ot the species here are primarily blue in males, brown without rufous breast in females (with notable exception of White-bellied). Note overall tone, pattern of face and vent.

126.1. Verditer Flycatcher
Eumyias thalassinus

15cm p.381
Slim, small-headed and long-tailed, with short black bill and sharp white scales on vent. Male is vivid turquoise with black lores; female greyer. See Black-naped Blue Monarch (Pl. 110).

126.2. Dusky Blue Flycatcher
Eumyias sordidus

14cm p.381
Small, short-tailed and chunky; ashy-blue with dark bill, lores and legs, and whitish belly.

126.3. Pale Blue Flycatcher
Cyornis unicolor

16cm p.384
Mid-sized with large dark bill. Male blue (not greenish) with greyer belly and weakly scaled vent. Female is olive-brown above with pale eye-ring, rufous rump and tail, and greyish below with whitish throat and belly.

126.4. Nilgiri Flycatcher
Eumyias albicaudatus

15cm p.381
Dark, short-billed, with small white basal panels on tail (very hard to see). Male dark steely-blue with brighter forehead and dark lores. Female grey with brownish upperparts and blue tint to wings and tail.

126.5. White-bellied Blue Flycatcher
Cyornis pallipes

15cm p.384
Longish-billed and pale-legged, with grey flanks, whitish belly and vent. Male cobalt-blue (not dark or greenish) with brighter forehead and brow. See similar but darker and much longer-legged sympatric White-bellied Blue Robin (Pl. 115). Female is olive-brown with greyish face, whitish throat and spectacles, and chestnut breast (darker than in similar female 'blue-and-rufous' flycatchers, Pl. 125).

126.6. Small Niltava
Niltava macgrigoriae

11cm p.382
Much smaller than Large. Male is blue (*vs* black) with brighter blue accents than male Large, and grey belly. Female is plainer than female Large, with no grey hindcrown or buff throat-streak.

126.7. Large Niltava
Niltava grandis

20cm p.382
Very large with long rounded tail, and bright blue neck-spot. Male black with deep blue crown, wings, rump and tail; female olive-brown with greyish hindcrown, rufescent face, wings, rump and tail, and contrasty buff throat-streak. Similar female Vivid (Pl. 125) is greyer-headed with more rufescent breast.

126.8. White-tailed Blue Flycatcher
Cyornis concretus

18cm p.383
Large and heavy-billed, with white slots in outertail feathers, and white belly. Male deep blue with bright blue brow and grey flanks (approaching allopatric White-bellied). Female is olive-brown with buffy spectacles and throat, and white gorget; see female Rufous-bellied Niltava (Pl. 125).

Plate sponsored by Mr. Jackson Burke.

PLATE 127. LAUGHINGTHRUSHES I (J. Anderton)

Laughingthrushes (formerly united in *Garrulax*; Plates 127–130)

Gregarious, rather jay-like babblers, many of which are very noisy but often elusive and hard to see. Usually in flocks, often single-species. Rarely colourfully patterned. Note eye colour, presence of contrasting tail-tips, head pattern.

127.1. Streaked Laughingthrush
Trochalopteron lineatum

20cm p.415
Small and stubby-billed, grey with fine (mostly white) streaks, and rufescent ear-coverts, wings and grey-tipped tail; more rufous in E.

127.6. Assam Laughingthrush
Trochalopteron chrysopterum

28cm p.417
From allopatric Red-headed by grey forehead and brow, plainer ear-coverts, and dark chestnut throat.

127.2. Bhutan Laughingthrush
Trochalopteron imbricatum

20cm p.415
Browner than Streaked with greyish lores and cheek, weaker shaft-streaks, narrower grey tail-tips and no rich rufous in plumage.

127.7. Red-headed Laughingthrush
Trochalopteron erythrocephalum

28cm p.417
Golden-olive wing- and tail-panels, chestnut nape, pale-scaled cheek, dark eye and black chin and spots on mantle and breast; pale grey-brown with solid rufous crown in W and C Himalayas, more chestnut with grey crown and silvery cheeks in E.

127.3. Blue-winged Laughingthrush
Trochalopteron squamatum

25cm p.416
Brown and heavily scaled black. Note black crown-stripe and tail with rufous wings, rump and tail-tip; eye often white.

127.8. Rufous-chinned Laughingthrush
Ianthocincla rufogularis

22cm p.413
Heavily spotted black overall, with black crown, blotchy moustachial, wing-slash and tail-band, and prominent buff lorespot; pale throat with variable amount of rufous. Note rufous vent and tail-tip.

127.4. Brown-capped Laughingthrush
Ianthocincla austeni

22cm p.413
Maroon-brown with white scales below, whitish wing-bar and tail-tips.

127.9. Ashy Laughingthrush
Ianthocincla cineracea

22cm p.412
Buff-brown with pale eye, and black crown, eyeline and moustache; note also black carpal patch and subterminal band on wing and tail, with white tips.

127.5. Scaly Laughingthrush
Trochalopteron subunicolor

23cm p.416
Olive and heavily scaled with extensive green-gold in wing and tail, slaty head, white eye (usually) and tail-tip.

PLATE 128. BABAXES AND LAUGHINGTHRUSHES II (J. Anderton)

Babaxes *Babax*

Heavily streaked babblers with heavy, curved blackish bills, and whitish eyes on rather pale faces.

128.1. Mount Victoria Babax
Babax [lanceolatus] woodi

28cm p.447
Smaller and paler than Giant, with brownish tail, broader dark moustache and conspicuous pale face.

128.3. Giant Babax
Babax waddelli

31cm p.447
Hypothetical; Sikkim. Larger than Mount Victoria and greyer especially on tail, with longer bill.

128.2. Tibetan Babax
Babax koslowi

p.448
Possible; extreme NE (not illustrated); more rufous and less streaked than congeners.

Laughingthrushes (continued from Plate 127 and on Plates 129–130), Liocichla

128.4. Spot-breasted Laughingthrush
Stactocichla merulina

22cm p.412
Plain with rufous vent, fine white postocular brow, pale eye, black streaks on pale throat and breast.

128.5. Red-faced Liocichla
Liocichla phoenicea

23cm p.418
Dark brown with dark red cheek and wing-slash, and rufous-tipped black tail.

128.6. Rufous-necked Laughingthrush
Dryonastes ruficollis

23cm p.411
Dark with slaty crown, black face and breast, and rufous neck-patch and vent.

128.7. Black-chinned Laughingthrush
Trochalopteron cachinnans

20cm p.414
Like Kerala, but with black eyeline and throat, and slatier-grey crown; note longer frilly white brow. In Nilgiris (*cachinnans*) cheek to breast are rufous, grey farther NW (*jerdoni*).

128.8. Kerala Laughingthrush
Trochalopteron fairbanki

20cm p.414
Small, with greyish-brown crown and eye-stripe, white brow, greyish throat and breast, and rufous flanks. In extreme S (*meridionale*), has paler brown crown, shorter white brow, greyer upperparts, more streaked breast, and whiter belly centre than nominate of farther north.

128.9. Ashy-headed Laughingthrush
Garrulax cinereifrons

23cm p.409
Only laughingthrush in Sri Lanka; uniform rufous-brown with grey crown, dark bill and legs, and pale eyes. See Ceylon and Indian Rufous Babblers (Pl. 132).

128.10. Spotted Laughingthrush
Ianthocincla ocellata

32cm p.413
Large with white tail-tips, chestnut wings, tail and face, black-and-white spots on mantle, and black throat and breast.

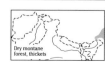

128.11. Giant Laughingthrush
Ianthocincla maxima

35cm p.413
Possible; N Arunachal. Very like Spotted but with chestnut throat and buff breast; longer tail is grey.

PLATE 129. LAUGHINGTHRUSHES III (J. Anderton)

Laughingthrushes (continued from Plates 127–128 and on Plate 130)

129.1. Elliot's Laughingthrush
Trochalopteron elliotii
25cm p.416
Hypothetical; NE Arunachal. Greyish-brown with blackish face, pale forehead and pale eye, chestnut vent and golden fringes on wings and tail. See Great Parrotbill (Pl. 140).

129.2. Black-faced Laughingthrush
Trochalopteron affine
25cm p.417
Blackish head with white moustache and post-auricular patches, golden fringes on bluish wings and tail.

129.3. Variegated Laughingthrush
Trochalopteron variegatum
24cm p.414
Pale brown with black mask and throat, pale moustachial to malar, and black tail with broad greyish subterminal band and white tip. Wing/tail edges silver in NW, yellow farther E.

129.4. Brown-cheeked Laughingthrush
Trochalopteron henrici
25cm p.417
Hypothetical; NE. Grey with white tail-tips, brown cheek and fine white brow and submoustachial.

129.5. White-browed Laughingthrush
Dryonastes sannio
23cm p.411
Brown with buffy-white patch on brow, lores and cheek.

129.6. Striped Laughingthrush
Trochalopteron virgatum
25cm p.416
Slim, rufescent, with narrow white streaks, broad whitish brow, cheek to malar.

129.7. Yellow-throated Laughingthrush
Dryonastes galbanus
23cm p.411
Smaller than Rufous-vented, and paler above with grey tail with broad black tip, black mask reaching chin, yellow belly.

129.8. Chestnut-backed Laughingthrush
Dryonastes nuchalis
23cm p.410
Dark with black face and throat, grey crown, large white ear-patch and rufous nape.

129.9. Rufous-vented Laughingthrush
Dryonastes gularis
23cm p.411
Like Wynaad but paler above, with yellow throat, black bill, rufous outertail.

129.10. Wynaad Laughingthrush
Dryonastes delesserti
23cm p.411
Dark with slaty crown, black mask and white throat, grey breast and rufous belly.

PLATE 130. LAUGHINGTHRUSHES IV (J. Anderton)

Laughingthrushes (continued from Plates 127–129)

130.1. Striated Laughingthrush
Grammatoptila striata

28cm   p.410
Large, stout-billed with bushy chestnut crest; body entirely streaked white. Black crown-stripe in NE.

130.5. White-crested Laughingthrush
Garrulax leucolophus

28cm   p.409
Dark brown with white hood and crest and black mask.

130.2. White-throated Laughingthrush
Garrulax albogularis

28cm   p.409
White throat, buffy belly and vent, and white tips to outertail.

130.6. Greater Necklaced Laughingthrush
Garrulax pectoralis

29cm   p.410
This and Lesser share white outer tail-tips, bold black necklace. Larger than Lesser with dark eye and yellow eye-ring, pale lores, complete black moustachial line from gape to necklace, (usually) buffy throat, and blackish primary coverts. Cheek may be white, streaked or solid black.

130.3. Grey-sided Laughingthrush
Dryonastes caerulatus (part)

25cm   p.412
Reddish-brown above and white below with broadly grey flanks, black face, black scales on crown and white ear-patch.

130.7. Lesser Necklaced Laughingthrush
Garrulax moniliger

27cm   p.409
Smaller than Greater with pale eye, usually 'open' cheek-patch, and whiter throat with rufous border to necklace; dark eye-ring.

130.4. *Dryonastes caerulatus subcaerulatus*

Entire cheek white in Meghalaya, with broad white tail-spots.

PLATE 131. GRASSLAND BABBLERS, TIT-BABBLER AND *STACHYRIS* (J. Anderton)

Grassland babblers *Chrysomma, Dumetia, Rhopocichla, Timalia*

131.1. Jerdon's Babbler
Chrysomma altirostre
17cm p.442
Like Yellow-eyed but paler-billed, greyer-throated, with grey brow and dark eye.

131.3. Tawny-bellied Babbler
Dumetia hyperythra
13cm p.440
Small and longish-tailed with rufous crown and underparts; in S, throat and brow white.

131.2. Yellow-eyed Babbler
Chrysomma sinense
18cm p.441
Long-tailed with stout black bill, white brow, throat and breast, yellow eye and eye-ring.

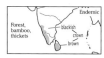

131.4. Dark-fronted Babbler
Rhopocichla atriceps
13cm p.441
Compact and short-tailed, with black cap, pale eyes, and whitish underparts. Crown brown in Sri Lanka.

131.5. Chestnut-capped Babbler
Timalia pileata
17cm p.441
Dark with chestnut crown, white face and throat with broad black eyeline and bill.

Babblers *Macronous, Stachyris*

Small species mostly of subdued coloration, longer-billed than fulvettas (*Alcippe*, Pl. 135) and not as tit-like. These *Stachyris* have strikingly pointed bills. Note pattern of head, colour of crown.

131.6. Striped Tit-babbler
Macronous gularis
11cm p.441
Olive above with rufous crown, wings and tail, and finely streaked yellow underparts; note yellow brow and pale eye.

131.9. Golden Babbler
Stachyris chrysaea
10cm p.439
Bright yellow-orange head and underparts, with black-streaked crown, black lores. In Mizoram, has contrastingly greyish cheek.

131.7. Rufous-capped Babbler
Stachyris ruficeps
12cm p.439
Like Rufous-fronted but yellower overall, with brighter cap and no greyish brow; bill a bit bigger and pinker.

131.10. Grey-throated Babbler
Stachyris nigriceps
12cm p.439
Buff-brown with white brow, eye-ring and submoustachial, blackish throat and white streaks on black crown.

131.8. Rufous-fronted Babbler
Stachyris rufifrons
12cm p.438
Like Rufous-capped but buffier, lacking yellow cast, with greyish lores and brow, duller rufous crown, and smaller, darker bill.

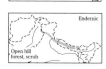

131.11. Black-chinned Babbler
Stachyris pyrrhops
10cm p.439
Buffy with black lores and throat; red eyes.

PLATE 132. *TURDOIDES* BABBLERS (J. Anderton)

Babblers *Turdoides*

Longish-tailed, (mostly) pale-eyed babblers, many very drab and unkempt. Extremely gregarious and noisy, with distinctive nasal, wheezy or bleating voices. Note degree of streaking on upper- and underparts, bill colour.

132.1. White-throated Babbler
Turdoides gularis

p.444
Possible; extreme SE (not illustrated). Recalls Common but throat and upper breast contrastingly white.

132.7. Spiny Babbler
Turdoides nipalensis

25cm   p.443
Brown above with curved blackish bill, variably whitish face, and sharply streaked underparts.

132.2. Afghan Babbler
Turdoides huttoni

>23cm   p.443
Larger, paler and greyer than Common, with weaker streaks on mantle but heavier, more distinct ones on breast, and without rufous tone to cheek and flanks.

132.8. Jungle Babbler
Turdoides striata

25cm   p.445
Very drab grey with yellowish bill; mottled below in S, paler in NW, and unmottled below in N. N W Ghats form *somervillei* has rufous tail and buff belly, with dark brown primaries.

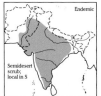

132.3. Common Babbler
Turdoides caudata

<23cm   p.443
Pale with streaked upperparts, pinkish-buff cheek and flanks. Compare with large grass warblers (Pl. 141).

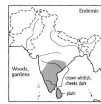

132.9. Yellow-billed Babbler
Turdoides affinis

23cm   p.447
Two very different forms, both with pale grey panel in flight feathers. In Peninsular nominate, crown variably whitish, breast heavily mottled and mantle streaked, rump pale grey and tail with broad blackish tip. Sri Lankan *taprobanus* nearly solid pale grey with paler wing-panel; no overlap with very similar Jungle.

132.4. Slender-billed Babbler
Turdoides longirostris

23cm   p.444
Unstreaked with curved black bill; rufous above with whitish face and buff underparts.

132.10. Indian Rufous Babbler
Turdoides subrufa

25cm   p.444
Entirely rufous below, with silvery forecrown, black-and-yellow bill and black lores.

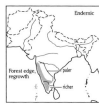

132.5. Striated Babbler
Turdoides earlei

20cm   p.444
Dusky, streaked above and below, with rufescent underparts.

132.11. Ceylon Rufous Babbler
Turdoides rufescens

25cm   p.447
Chestnut with orange bill and legs, and ashy crown and nape. Ashy-headed Laughingthrush (Pl. 128) has black bill.

132.6. Large Grey Babbler
Turdoides malcolmi

28cm   p.444
Large and long-tailed, with stout bill, black lores, silvery forehead and creamy-grey outertail.

PLATE 133. SCIMITAR-BABBLERS (J. Anderton)

Scimitar-babblers *Xiphirhynchus, Pomatorhinus*

Handsome babblers with longish, decurved bills and bold plumage patterns, usually found feeding low in dense thickets and forest. Most species have distinctive liquid, piping calls. Note pattern of underparts, presence of supercilium, bill and eye colour.

133.1. Slender-billed Scimitar-babbler
Xiphirhynchus superciliaris
20cm p.430
Very long curved bill; dusky rufous-brown with dark grey head and whitish brow and throat.

133.2. Streak-breasted Scimitar-babbler
Pomatorhinus ruficollis
19cm p.429
Smaller, shorter-billed than White-browed with streaked breast, brown crown and dark eye.

133.3. Ceylon Scimitar-babbler
Pomatorhinus [schisticeps] melanurus
<22cm p.429
Smaller than Indian, with smaller, less curved bill and chestnut mantle and flanks, brightest in wet zone.

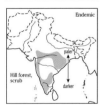

133.4. Indian Scimitar-babbler
Pomatorhinus [schisticeps] horsfieldii
>22cm p.428
Cold brown with long, curved yellow bill, long white brow and bib; no rufous on neck.

133.5. White-browed Scimitar-babbler
Pomatorhinus schisticeps
22cm p.429
From Indian by grey crown, rufous neck-sides and flanks, and pale eye.

133.6. Large Scimitar-babbler
Pomatorhinus hypoleucos
28cm p.427
Dark with long pale bill, rufous neck, and greyish cheeks and flanks; no white brow.

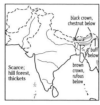

133.7. Coral-billed Scimitar-babbler
Pomatorhinus ferruginosus
22cm p.430
Mid-length bill red, mask black, brow and chin-strap white. In E Himalayas, crown black and underparts chestnut. In S Assam hills, paler chestnut to buff below, with brown crown.

133.8. Spot-breasted Scimitar-babbler
Pomatorhinus erythrocnemis
22cm p.427
Like Rusty-cheeked but breast spotted, flanks olive with rufous only on cheek and vent.

133.9. Rusty-cheeked Scimitar-babbler
Pomatorhinus erythrogenys
25cm p.428
Rufous face, flanks and vent, with whitish throat and breast; no pale brow.

133.10. Long-billed Scimitar-babbler
Pomatorhinus ochraceiceps
23cm p.429
From NE races of Coral-billed by longer, slimmer orange bill, rustier crown with no black border above white brow.

PLATE 134. MYZORNIS, MESIAS AND YUHINAS (J. Anderton)

Myzornis *Myzornis*

134.1. Fire-tailed Myzornis
Myzornis pyrrhoura
12cm   p.461
Sunbird-like; mossy green with scaly crown and red slashes in wings and tail; red throat and vent brighter on male.

Mesia, Leiothrix *Leiothrix*

Rather small, brightly coloured, gregarious babblers with musical voices; males brightest, juveniles dullest.

134.2. Silver-eared Mesia
Leiothrix argentauris
15cm   p.448
Olive above and orange-yellow below with yellow bill, black crown and tail, silvery cheeks. See Golden-breasted Fulvetta (Pl. 135).

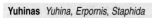

134.3. Red-billed Leiothrix
Leiothrix lutea
13cm   p.448
Olive with red bill, yellow throat, whitish face with black whisker-mark, notched black tail.

Yuhinas *Yuhina, Erpornis, Staphida*

A heterogeneous assemblage of small, crested species, most of which are rather sombrely coloured. Note colour of nape and throat, and facial pattern, crest shape.

134.4. Black-chinned Yuhina
Yuhina nigrimenta
11cm   p.460
Tiny. Grey above and white below with grey head, and black crest and face; red bill.

134.5. White-bellied Erpornis
Erpornis zantholeuca
11cm   p.467
Yellowish-olive with short crest, pinkish bill, greyish face, and bright yellow vent; see Yellow-browed Tit (Pl.156), Green Shrike-babbler (Pl. 137).

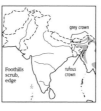

134.6. Striated Yuhina
Staphida castaniceps
13cm   p.458
Stubby bill, rounded tail with white outer tips, white shaft-streaks on olive-brown mantle, and rufous cheek-patch. Short crest is grey in Himalayas, rufous with grey-scaled forecrown in S Assam hills.

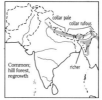

134.7. Whiskered Yuhina
Yuhina flavicollis
13cm   p.459
Grey head with blackish crest and whisker-mark, white eye-ring and throat, yellowish hindcollar varying from whitish in W to rufous in E.

134.8. White-naped Yuhina
Yuhina bakeri
13cm   p.459
Brown with crested rufous head, white nape-patch and streaks on cheek.

134.9. Rufous-vented Yuhina
Yuhina occipitalis
13cm   p.459
Crested grey head with rufous nape and white eye-ring; rufous vent.

134.10. Stripe-throated Yuhina
Yuhina gularis
14cm   p.459
Large and dusky with slim crest, black streaks on pale throat, and orange-buff wing-slash.

PLATE 135. FULVETTAS (J. Anderton)

Fulvettas or tit-babblers *Alcippe*

Small, big-headed, active babblers, often in mixed feeding flocks, usually in undergrowth and low forest-levels; very tit-like. Most are primarily brown with contrasting wing-slashes. Note wing and head patterns.

135.1. Rufous-winged Fulvetta
Alcippe castaneceps

10cm   p.453
Olive; rufous crown and wing-slash, white face with blackish eye-stripe and whisker-mark.

135.6. Brown-throated Fulvetta
Alcippe ludlowi

11cm   p.454
From White-browed by lack of white brow; note also golden edges of tertials, grey and chestnut underparts.

135.2. Yellow-throated Fulvetta
Alcippe cinerea

10cm   p.453
Greyish-olive with yellow head and underparts, black eyeline, lateral crown-stripe and scales on crown. See leaf-warblers (Pl. 152), Black-faced Warbler (Pl.149).

135.7. Manipur Fulvetta
Alcippe manipurensis

11cm   p.454
Pale brownish-grey head with dark lateral crown-stripe and streaks on throat; otherwise similar to White-browed.

135.3. Golden-breasted Fulvetta
Alcippe chrysotis

11cm   p.452
Grey with yellow underparts, wing- and tail-slashes, and blackish hood with silvery ear-patch; white coronal stripe in S Assam hills. See Silver-eared Mesia (Pl. 134).

135.8. Rusty-capped Fulvetta
Alcippe dubia

13cm   p.454
Large, longish-tailed and dark with white postorbital brow, rufous crown with black scales and lateral stripe, dark face and streaked neck; buffy below.

135.4. White-browed Fulvetta
Alcippe vinipectus

11cm   p.453
Brown above with dark crown and mask, white brow and throat, silver and black slashes in rufous wing. E races have fine spots on throat, coarser in S Assam hills.

135.9. Rufous-throated Fulvetta
Alcippe rufogularis

12cm   p.454
Rufous crown and half-collar, black lateral crown-stripe, white brow, throat and breast. Facial pattern recalls Snowy-browed Babbler (Pl. 139).

135.5. Chinese Fulvetta
Alcippe striaticollis

11cm   p.454
Possible in NE; note dark streaks on crown, blackish lores, sharp streaks on whitish throat.

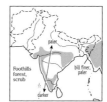

135.10. Brown-cheeked Fulvetta
Alcippe poioicephala

15cm   p.455
Plain with grey head, browner on cheek and whitish on throat; no crown-stripes. In Peninsula, bill thick and all-black; in NE, bill finer, more decurved, paler.

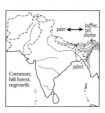

135.11. Nepal Fulvetta
Alcippe nipalensis

12cm   p.455
From Brown-cheeked by bold white eye-ring, blackish brow; see Blue-winged Minla (Pl. 137).

PLATE 136. BARWINGS, SIBIAS AND WHITE-HOODED BABBLER (J. Anderton)

Barwings *Actinodura*

Medium-sized brown babblers with loose shaggy crests and densely barred wings and tails. Acrobatic feeders, often clinging to tree-trunks to forage in patches of moss. Note pattern of head and throat, degree of tail-barring.

136.1. Rusty-fronted Barwing
Actinodura egertoni

23cm   p.449
Longish tail lacks black tip; note loose grey crest, chestnut face and pale bill; buffy to rufous below.

136.2. Hoary-throated Barwing
Actinodura nipalensis

20cm   p.450
Black moustache borders unstreaked pale grey throat and breast; note pale-streaked high crest.

136.3. Streak-throated Barwing
Actinodura waldeni

20cm   p.450
From Hoary-throated by blurry brown streaks on throat and breast, pale-scaled crest, and much weaker dark moustache. Throat is rufescent in S Assam hills, greyish in NE (where very like Hoary-throated; see text).

Sibias *Heterophasia*

Bulbul-sized, showy, slim babblers with long narrow tails and slim pointed bills; most are at least slightly crested. Arboreal and noisy. Note colour of throat and back, tail pattern.

136.4. Rufous-backed Sibia
Heterophasia annectans

18cm   p.457
Small; mostly black above and white below with rufous back, rump and flanks and white-tipped black tail.

136.5. Rufous Sibia
Heterophasia capistrata

21cm   p.457
Rufous with black cap, black-and-grey wings, tail banded terminally black and grey.

136.6. Grey Sibia
Heterophasia gracilis

21cm   p.457
Grey above and white below with black cap and peach-washed belly and vent; tail and wing-patterns like Rufous.

136.7. Long-tailed Sibia
Heterophasia picaoides

30cm   p.458
All grey with very long, graduated white-tipped tail and white wing-slash. See Green-billed Malkoha (Pl. 74) and blue magpies (Pl. 180).

136.8. Beautiful Sibia
Heterophasia pulchella

22cm   p.457
Slate-grey with blackish face and tail with broad grey terminal band.

White-hooded Babbler *Gampsorhynchus*

136.9. Indian White-hooded Babbler
Gampsorhynchus rufulus

23cm   p.449
White head and underparts, pale bill, legs and eye; tail with white outer tips. Juv. has rufous cap and buffier underparts; see rufous-headed parrotbills (Pl. 140).

Plate sponsored by Mrs. Pamela Copeland.

PLATE 137. MINLAS AND SHRIKE-BABBLERS (J. Anderton)

Minlas *Minla*

Colourful but dissimilar species, often found in treetop feeding parties. Active and highly gregarious, but not very vocal.

137.1. Blue-winged Minla
Minla cyanouroptera

15cm p.450
Slim and pale with blue wings and tail, blue-grey crown and whitish face and underparts.

137.2. Bar-throated Minla
Minla strigula

14cm p.451
Fulvous-olive with ragged golden-rufous crest, black whisker-mark and bars on throat, and yellow belly; orange-yellow, black and whitish patches in wings. See barwings (Pl. 136).

137.3. Red-tailed Minla
Minla ignotincta

14cm p.451
Bold black crown and mask with white brow and throat, pale eye; red trim on black wings and tail. See White-browed Shrike-babbler.

Shrike-babblers *Pteruthius*

Stout, short-tailed, rather sluggish babblers with heavy, hooked bills. Note especially head and wing patterns.

137.4. Green Shrike-babbler
Pteruthius xanthochlorus (part)

13cm p.467
Olive-drab with round grey head (darker in male), small stubby bill and slight wing-bar. See leaf warblers (Pl. 151, 152) and Yellow-browed Tit (Pl. 156).

137.5. *Pteruthius xanthochlorus hybrida*

13cm
Like N races but eye-ring white.

137.6. Black-eared Shrike-babbler
Pteruthius melanotis

11cm p.467
Olive above and yellow below with black crescent on cheek. Male has double white wing-bar and diffuse chestnut throat; female is buffy on wing-bars, forehead and throat.

137.7. Chestnut-fronted Shrike-babbler
Pteruthius aenobarbus

11cm p.467
Single regional specimen. From Black-eared by lack of black cheek-crescent, and olive vs grey wings. Male has dark rufous above bill; female is whitish below with rufous forehead and chin.

137.8. White-browed Shrike-babbler
Pteruthius flaviscapis

16cm p.465
Chestnut tertials and white throat. Male is grey above with black cap, wings and tail, white postocular brow; female is grey-crowned with olive mantle and yellow-green wings.

137.9. Black-headed Shrike-babbler
Pteruthius rufiventer

17cm p.465
Grey throat and breast, and rufous belly, rump and tail-tips. Male has black cap, wings and tail, and chestnut mantle; female has grey head-sides, olive-green mantle.

Cutia *Cutia*

137.10. Cutia
Cutia nipalensis

20cm p.418
Heavy hooked bill and dark mask, black-barred flanks, and rufous rump and tail-coverts; back in male rufous, brown with black streaks in female.

PLATE 138. WREN AND WREN-BABBLERS I (J. Anderton)

Wren Troglodytidae

Tiny, noisy, skulking wren-babbler-like bird. The only regional member of its family.

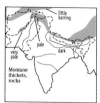

138.1. Winter Wren
Troglodytes troglodytes

9cm p.353
Recalls wren-babblers, with short cocked tail, but finely barred on wings, tail and flanks (at least). Extreme regional colour variation.

Wren-babblers *Pnoepyga, Spelaeornis*

Pnoepyga is virtually tailless, and underparts are either mainly white or fulvous in adults, solid dark brown in juvs.

138.2. Scaly-breasted Wren-babbler
Pnoepyga albiventer

10.5cm p.434
Relatively large and scaly, with buff spots on neck (sometimes crown and upper mantle and wing-coverts).

138.3. Pygmy Wren-babbler
Pnoepyga pusilla

9cm p.434
Tiny and scaly; has pale spots on wing-coverts and tertials, but not about head.

138.4. Nepal Wren-babbler
Pnoepyga immaculata

10cm p.434
Like Scaly-breasted and Pygmy, but narrowly streaked on throat and breast, with no white/buff spots above; bill longer.

138.5. Mishmi Wren-babbler
Spelaeornis badeigularis

9cm p.435
From Rufous-throated by much darker throat with black streaks, darker grey face and whitish chin.

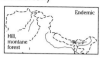

138.6. Rufous-throated Wren-babbler
Spelaeornis caudatus

9cm p.434
Short but visible tail; unmarked rufous throat and densely black-and-white-barred belly.

138.7. Spotted Wren-babbler
Spelaeornis formosus

10cm p.435
Brown, peppered with white spots; note heavily black-barred rufous wing and short tail. Juvenile much darker with bigger white spots.

138.8. Chin Hills Wren-babbler
Spelaeornis oatesi

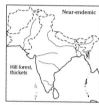

10cm p.436
Heavily scaled above with white throat and breast densely speckled with black.

138.9. Naga Wren-babbler
Spelaeornis chocolatinus

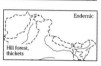

10cm p.435
From Tawny-breasted by darker upperparts, and fine black speckles on breast; more rufescent below than Grey-bellied, with stronger pale streaks on flanks.

138.10. Bar-winged Wren-babbler
Spelaeornis troglodytoides

10cm p.435
Dark rufous with bright white bib and barred wings and tail; note dark crown with pale spots. Richer colour in SE Arunachal, where wings barred black-and-white.

138.11. Tawny-breasted Wren-babbler
Spelaeornis longicaudatus

11cm p.436
Relatively plain with fulvous, almost unmarked underparts and weakly scaled upperparts.

138.12. Grey-bellied Wren-babbler
Spelaeornis reptatus

10cm p.436
Dark and greyish; from Tawny-breasted by streaked flanks; browner below than Naga. See text. Male greyish below with brown streaks.

PLATE 139. WREN-BABBLERS II AND FOREST-FLOOR BABBLERS (J. Anderton)

Wren-babblers *Napothera, Rimator*

139.1. Eyebrowed Wren-babbler
Napothera epilepidota
10cm p.433
Heavily scaled above with buffy brow, white spots on wing-coverts; entirely streaked below.

139.2. Streaked Wren-babbler
Napothera brevicaudata
12cm p.433
Dark, and coarsely streaked above with greyish face; no pale brow or streaks on flanks.

139.3. Long-billed Wren-babbler
Rimator malacoptilus
12cm p.433
Long, fine bill; heavily streaked buff overall with dark whisker-mark.

Wedge-billed Babblers *Sphenocichla*

Mid-sized, dark, gregarious, little-known babblers. Formerly considered a single species of wren-babbler.

139.4. Cachar Wedge-billed Babbler
Sphenocichla roberti
18cm p.438
Wedge-shaped, pointed bill; dark brown above with grey scales, and densely scaled black-and-white below.

139.5. Sikkim Wedge-billed Babbler
Sphenocichla humei
17cm p.436
From Cachar by broad scaly pale grey brow behind eye, and black face and underparts.

Babblers *Pellorneum, Malacocincla, Stachyris*

Largely terrestrial; note head pattern, colour and pattern of underparts.

139.6. Buff-breasted Babbler
Pellorneum tickelli
15cm p.424
Olive-brown with buff underparts, including throat; rather featureless with squared, mid-length tail. Lacks grey on face.

139.7. Spot-throated Babbler
Pellorneum albiventre
14cm p.424
Smaller than Buff-breasted; much smaller and smaller-billed than Abbott's. In E Himalayas, from Buff-breasted by greyish face and shorter tail; in S Assam hills, more like Buff-breasted but throat whiter, tail rounder. Throat-spots can be obsolete; see Lesser Shortwing (Pl. 115).

139.8. Marsh Babbler
Pellorneum palustre
15cm p.424
Recalls Puff-throated, but note white eye-ring (no brow), and rufescent cheek, flanks and vent.

139.9. Abbott's Babbler
Malacocincla abbotti
15cm p.424
Stout with heavy bill and short tail; drab with diffuse grey throat and brow, rufescent vent.

139.10. Snowy-throated Babbler
Stachyris oglei
13cm p.440
Chunky with heavy black bill, black mask, clean white brow and throat, streaked neck.

139.11. Brown-capped Babbler
Pellorneum fuscocapillus
16cm p.427
Plain with blackish crown and cinnamon-buff underparts.

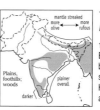

139.12. Puff-throated Babbler
Pellorneum ruficeps
15cm p.425
Brown above with rufescent crown and broad pale brow; white below with strongly streaked breast and flanks. Himalayan forms scaled on forehead and side of neck.

PLATE 140. PARROTBILLS (J. Anderton)

Parrotbills *Paradoxornis, Panurus, Conostoma*

Peculiar, rather long-tailed babblers; most have very deep, short bills designed for cutting bamboo. The larger species are retiring and difficult to observe; the smaller are hyperactive. Note throat and face pattern.

140.1. Fulvous Parrotbill
Paradoxornis fulvifrons

12cm   p.463
Small; buff with rufous wings and tail, greyish crown-stripes. Compare Rufous-fronted Tit (see also Pl. 155).

140.2. Black-throated Parrotbill
Paradoxornis nipalensis

10cm   p.463
Tiny with a (usually) tiny black bill; largely orange-rufous above with white face, black throat and brow. Variable in colour of crown, cheek and breast.

140.3. Grey-headed Parrotbill
Paradoxornis gularis

16cm   p.462
Rather plain with yellow bill, grey head with black crown-stripe and throat.

140.4. Bearded Reedling
Panurus biarmicus

15cm   p.461
Pale with long graduated tail and small yellow bill; rufous above with black markings, white primaries and greyish head. Male has black vertical face-patch; juv. very distinctive.

140.5. Brown Parrotbill
Paradoxornis unicolor

21cm   p.462
Large; greyish-brown with stubby yellowish bill and coarsely streaked head.

140.6. Lesser Rufous-headed Parrotbill
Paradoxornis atrosuperciliaris

15cm   p.463
Smallish with pink bill; olive-brown above with rufous rounded head, short black eyebrow, paler cheeks, pale blue lores and eye-ring. Rare Himalayan race lacks eyebrow.

140.7. Greater Rufous-headed Parrotbill
Paradoxornis ruficeps

18 cm   p.463
Larger than Lesser with uniform rufous cheeks and no black brow.

140.8. Great Parrotbill
Conostoma oemodium

28cm   p.461
Very large with thick stubby yellow bill; greyish with pale forehead and pale eye on blackish face. (Full figure at reduced scale.)

140.9. Black-breasted Parrotbill
Paradoxornis flavirostris

19cm   p.462
Large with steep yellow bill, large black ear- and throat-patches, and dark buff underparts.

140.10. Spot-breasted Parrotbill
Paradoxornis guttaticollis

19cm   p.462
Like Black-breasted but paler below with blurred black spots on whitish throat and breast.

PLATE 141. CISTICOLAS, *LOCUSTELLA* WARBLERS AND GRASSBIRDS (J. Anderton)

Cisticolas *Cisticola*

Tiny grassland warblers, heavily streaked above and unmarked below, with marked seasonal changes: tail much shorter in breeding plumage, but retaining diagnostic tail-tip patterns. Note colour of nape and rump, colour and pattern of crown and tail.

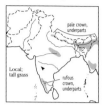

141.1. Bright-capped Cisticola
Cisticola exilis

10cm   p.468

From Zitting by less rufous rump, unstreaked hindcollar and dark tail with narrow pale (brown to grey) tips and large black subterminal bands; darker above, so streaks on mantle less bold. Female and non-breeding male have streaked crown and golden-buff brow. Breeding male has golden-orange cap and orange-buff underparts in S; cream-white cap and buff underparts in NE.

141.2. Zitting Cisticola
Cisticola juncidis

10cm   p.468

Paler above than Bright-capped, with bolder streaks on mantle, whitish brow, and diagnostic streaks on nape; note rufous rump and pale tail with broad white tips. Breeder has unstreaked dull pale brown crown, streaked in non-breeding.

Grasshopper-warblers *Locustella*

Very skulking winter visitors with short rounded tails. From reed- and bush-warblers (*Bradypterus*) by their streaked upperparts. Note streaking on breast and vent, facial pattern, rump colour.

141.3. Lanceolated Warbler
Locustella lanceolata

12cm   p.487

Small and short-tailed; darker olive-brown above than Grasshopper with streaks on crown and mantle (*vs* spots), and fine streaks on breast, flanks and vent; blackish tertials with narrow pale fringes when fresh. First-winter may be much less streaked below; from Grasshopper by finely streaked vent.

141.4. Grasshopper Warbler
Locustella naevia

13cm   p.488

Small; olive-grey with dark spots on crown and mantle, and pale (often yellowish) and almost unmarked below apart from fine arrowheads on vent (broader than in Lanceolated). Note dark tertials with broad olive-grey edges. First-winter can be like Lanceolated but vent streaks broader.

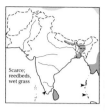

141.5. Rusty-rumped Warbler
Locustella certhiola

13cm   p.488

Large, streaked above, with rufous rump and white-tipped tail; long whitish brow over dark eye-stripe, and buffy below with plain vent. When worn, very dark above; first-winter yellowish and lightly streaked below.

141.6. Savi's Warbler
Locustella luscinioides

p.489

Possible: extreme NW (not illustrated). Unstreaked above, recalling European Reed, but bill shorter, undertail-coverts slightly pale-tipped.

Grassbirds *Graminicola, Megalurus, Chaetornis, Schoenicola*

141.7. Rufous-rumped Grassbird
Graminicola bengalensis

16cm   p.516

Broad dark tail tipped white below, strikingly streaked upperparts, rufous cheek, rump and wings; mostly white below.

141.8. Striated Grassbird
Megalurus palustris

male 25; female c. 22cm   p.515

Very large with long pointed tail and longish bill. Boldly streaked on mantle but not on rufescent crown (unlike Bristled), with pale brow. See *Turdoides* babblers (Pl. 132).

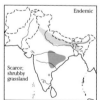

141.9. Bristled Grassbird
Chaetornis striata

20cm   p.515

Recalls Striated but bill stout, hindcrown streaked, tail broad and whitish-tipped. See *Turdoides* babblers (Pl. 132).

141.10. Indian Broad-tailed Grass-warbler
Schoenicola platyurus

18cm   p.516

Unstreaked brown with thick bill, heavy dark rounded tail, and very long, billowy, pale-tipped dark undertail-coverts.

PLATE 142. CONFUSING PRINIAS (J. Anderton)

Prinias *Prinia* (continued on Plate 143)

Mostly small, slim and long-tailed warblers with pronounced seasonal variation in plumage, including tail length (often much greater in non-breeding plumage) and sometimes tail pattern. Some species also exhibit significant but confusing geographic variation. The most similar and confusing unstreaked species shown here; most of them are also widespread. Note tail pattern and colour, bill size, and facial pattern.

142.1. Rufescent Prinia
Prinia rufescens

11cm p.472
Rufescent upperparts and tail, with buff or grey tail-tips. In breeder, short-tailed with grey head and whitish spectacles (lacking in Grey-crowned) and underparts; in non-breeder, long-tailed with dull rufous head, and weakly defined pale brow; from Grey-breasted and Plain by tail pattern.

142.2. Ashy Prinia
Prinia socialis

13cm p.473
Grey cap, rufescent wings and tail with whitish tips and black subterminal bands; rufescent underparts in most plumages. Darker above than Grey-breasted with heavier, longer bill, and never has greyish breast-band. Breeder usually lacks supercilium and has grey mantle; tail relatively short; non-breeder has fine short white brow and longer tail.

142.3. Grey-crowned Prinia
Prinia cinereocapilla

11cm p.471
Relatively short-tailed (in all plumages) with fine black bill, slaty crown with blackish lores, rufous-brown upperparts and tail, and rufescent tail-tips; buffy below. Non-breeder has broad buffy brow and somewhat longer tail.

142.4. Grey-breasted Prinia
Prinia hodgsonii

11cm p.472
Very like Rufescent in some plumages, but usually has at least some grey on breast, and has finer bill and greyer tail with white tips. Breeder has strong breast-band and grey upperparts and shortish tail; non-breeder has short whitish brow. Juv. rufescent above (especially on rump) and similar to Plain, but shorter-tailed and with a short white brow.

142.5. Jungle Prinia
Prinia sylvatica

15cm p.473
Larger than Plain, with heavier bill, weaker pale brow and heavier tail (broader at base); often small dark spots on malar. Breeding male has black bill, reduced brow, and largely white outertail. Great geographic variation; see text. Very large, heavy-set and short-tailed in Sri Lanka.

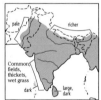

142.6. Plain Prinia
Prinia inornata

13cm p.476
Very similar to Jungle and equally variable, but smaller and more delicate, with narrower-based tail and finer bill; usually livelier. Drab greyish- to reddish-brown above, buffy-whitish below, with whitish lores and eyebrow. From similar plumages of Grey-breasted and Rufescent by weaker buffy tail-tips. Breeder has black bill, whitish outertail. See text. Short-tailed in Sri Lanka. Juv. Striated (Pl. 143.4) has streaked crown.

PLATE 143. COLOURFUL AND STREAKED PRINIAS (J. Anderton)

Scrub Warbler and **Prinias** *Scotocerca, Prinia* (continued from Plate 142)

Species on this plate are less variable and easier to identify than on previous plate, being either colourful or boldly streaked and/or patterned.

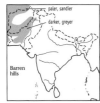

143.1. Scrub Warbler
Scotocerca inquieta

10cm   p.469
Pale grey with squared black tail, streaked crown and black eyeline, and cinnamon face and flanks.

143.6. Rufous-vented Prinia
Prinia burnesii

17cm   p.469
Grey-brown with long very broad tail, heavily streaked crown and mantle with pale cinnamon hindcollar, whitish spectacles, and unique rufous vent.

143.2. Rufous-fronted Prinia
Prinia buchanani

12cm   p.471
Pale brown with diagnostic large white tail-tips, rufous crown (subject to wear and lacking in juv.), and white brow, lores and underparts.

143.7. Graceful Prinia
Prinia gracilis

13cm   p.472
Very small with black-and-white-tipped tail, streaked crown and mantle, and white lores, eye-ring and underparts. Sandy-brown above in NW and NC; grey in NE and similar to much larger and broader-tailed Swamp.

143.3. Yellow-bellied Prinia
Prinia flaviventris

13cm   p.473
Olive above with grey head, white throat and breast, and yellow belly; short white brow in non-breeder. Drab in NW (extreme shown). Juv. is short-tailed and suffused yellow; see juv. *Cettia* bush-warblers (Pl. 144).

143.8. Swamp Prinia
Prinia cinerascens

16cm   p.470
Drab grey with mid-length broad tail, narrow white eye-ring and lore-spot, vaguely streaked upperparts, and whitish underparts. Much larger than very similar Graceful, which has a narrower tail and whitish tail-tips.

143.4. Striated Prinia
Prinia crinigera

16cm   p.470
Large and heavily streaked above with a very long graduated tail and rather heavy bill. Breeding male has blackish face, very dark upperparts and unstreaked underparts; non-breeding is paler-faced and finely streaked on breast. Juv. has shorter tail, diffuse streaks on crown and none on mantle; very similar sympatric Jungle (Pl. 142) has unstreaked crown.

143.9. Black-throated Prinia
Prinia atrogularis

17cm   p.470
Large and very long-tailed with unstreaked upperparts, and slaty cheek above whitish moustache; crown is grey-brown in Himalayas, rufous in S Assam hills. Breeder has black bib; non-breeder has some black streaks on throat and breast, and long white brow. Juv. is shorter-tailed and often has little black on throat.

143.5. Brown Prinia
Prinia polychroa

p.470
Hypothetical; possible in extreme SE (not illustrated). Like Striated but less streaked above than adult, paler and browner than NE Striated, with weaker dark tail-tips. Very similar to juvenile Striated.

143.10. Hill Prinia
Prinia superciliaris

p.471
Previously overlooked; extreme NE. Like Black-throated (Himalayan race) but never has black throat; supercilium more prominent and ear-coverts slatier.

PLATE 144. *CETTIA* BUSH-WARBLERS (J. Anderton)

Bush-warblers *Cettia*, *Urosphena*

Mostly skulking underbrush dwellers, some with extraordinary songs. Tails long or short, but not as broad and rounded as in *Bradypterus* (Pl. 145), but see Cetti's; rictal bristles prominent in most. Note face pattern, colour of upperparts, throat and flanks, tail length and shape.

144.1. Hume's Bush-warbler
Cettia brunnescens

11cm   p.483

Small, with fine dark bill and square-tipped tail. Fawn-brown above with rufescent wings and narrow whitish brow over narrow dark eye-stripe, and whitish or greyish below with pale yellowish belly. Shorter-tailed and browner above than 'Himalayan' Aberrant, with weaker brow; paler and more delicate than Strong-footed, with 'cleaner' underparts.

144.2. 'Himalayan' Aberrant Bush-warbler
Cettia flavolivacea (part)

13cm   p.482

Longish tail with rather round tip. Olive with dark eyeline, and yellowish brow and underparts. Longer-tailed than Hume's and Strong-footed, and much longer-tailed than Tickell's Leaf-warbler and similar *Phylloscopus* (Pl. 150).

144.3. 'Manipur' Aberrant Bush-warbler
Cettia flavolivacea weberi

Much more rufescent in S Assam hills, with bright cinnamon flanks on yellowish underparts; see Strong-footed.

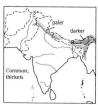

144.4. Strong-footed Bush-warbler
Cettia fortipes

12cm   p.481

Drab with mid-length tail, rather heavy dark bill and heavy legs; olive-grey to olive-brown above with pale brow and slightly rufescent wings, and slightly paler centre of throat and belly. Shorter-tailed and greyer (especially below) than Aberrant. Juv. very yellow below; browner above than 'Himalayan' Aberrant.

144.5. Large Bush-warbler
Cettia major

12cm   p.481

Short-tailed with a fine dark bill, chestnut crown and brownish flanks. Larger than Grey-sided, with brownish (*vs* grey) side of breast. See female Japanese; Pale-footed is much paler below and has olive-brown crown. Juv. very plain and dark.

144.6. Grey-sided Bush-warbler
Cettia brunnifrons

10cm   p.481

Small with bright rufous cap, pale supercilium and conspicuous grey upper flanks. Smaller than Large but longer-tailed, with brow better defined in front of eye. Juv. is much muddier and lacks rufous cap.

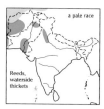

144.7. Cetti's Bush-warbler
Cettia cetti

13cm   p.483

Dull brown above with distinctive greyish brow and cheek, and white below with buffy undertail-coverts. Broad tail suggests a *Bradypterus* bush-warbler (Pl. 145).

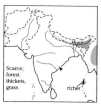

144.8. Pale-footed Bush-warbler
Cettia pallidipes

11cm   p.480

Small and 'clean' with shortish squared tail; brown above with buffy eyebrow over broad dark eye-stripe, and white below with pinkish-buff flanks and vent and pale pinkish legs. Shorter-legged and longer-tailed than similar Stub-tailed, shorter-tailed and paler below than any brownish leaf-warbler (Pl. 150).

144.9. Manchurian Bush-warbler
Cettia canturians

male 16cm; female 14cm   p.481

Large (especially male), long-tailed and heavy-billed; rufescent above with rufous crown, wings and tail. Male could be taken for Thick-billed Warbler (Pl. 146) but told by pale brow and blackish eyeline; smaller female from Large by much longer tail, heavier bill and less olive upperparts.

144.10. Stub-tailed Bush-warbler
Urosphena squameiceps

11cm   p.480

Hypothetical; E Nepal, NE. Chunky, very short-tailed and long-legged, with diagnostic black scales on crown. Otherwise similar to Pale-footed, with bolder, brighter buff brow and mottled cheek. Shape suggests a wren-babbler (Pl. 138–139) or a tesia (Pl. 148).

PLATE 145. *BRADYPTERUS* BUSH-WARBLERS (J. Anderton)

Bush-warblers *Bradypterus, Elaphrornis*

Elusive, drab warblers with simple songs; rictal bristles virtually lacking (unlike *Cettia*, Pl. 144). Most species have variable dark spots on throat and pale tips on undertail-coverts, with seasonally variable bill coloration. Tails short, fairly stiff and rounded. Note pattern of face and undertail-coverts, tone of upperparts.

145.1. Cetti's Bush-warbler
Cettia cetti See Pl.144, p.483

Rounded tail and general coloration suggest *Bradypterus*; whiter below with stronger supercilium.

145.2. Chinese Bush-warbler
Bradypterus tacsanowskius

14cm   p.485

Rather large and rather long-winged and -tailed, with heavier bill than similar *Bradypterus*. More olive above than similar Brown, with pale lores, narrow white broken eye-ring, often indistinct brow, and broad but indistinct pale tips to undertail-coverts; underparts whitish or yellowish with buffier flanks and sometimes fine specks or streaks on throat/breast. Bill dark in breeder, mostly pale on lower mandible in non-breeder. Juv. warmer brown above.

145.3. Baikal Bush-warbler
Bradypterus davidi   12.5cm   p.485

Hypothetical; NE. Smaller than very similar Spotted, and more olive above with little or no grey on brow and breast; breast-spots weaker than in most Spotted, but variable in both species. In winter, may have pale lower mandible.

145.4. Spotted Bush-warbler
Bradypterus thoracicus

13cm   p.485

Rather small with thin bill and dark brown undertail-coverts with broad, clear-cut white tips. Darker above than other *Bradypterus*; in breeder, has grey brow and breast, fine to heavy black spots on throat and breast, and black bill; in non-breeder, whiter on brow and below, with little or no spotting, and partly pale lower mandible. Juv. has yellowish underparts mottled brown (especially on breast), and weak buffy-yellow brow.

145.5. Brown Bush-warbler
Bradypterus luteoventris

14cm   p.486

Brown above with bright buffy brow, face, flanks and plain undertail-coverts, and whitish throat and centre of breast and belly. Lower mandible always pale. Paler above than Russet, darker than Chinese, and breast never spotted. Juv. darker above, yellowish below.

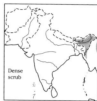

145.6. Russet Bush-warbler
Bradypterus mandelli

14cm   p.486

Similar to Brown, but heavier dark bill, darker chestnut-brown upperparts, and dark undertail-coverts with pale tips (less marked than in Spotted). Usually nearly lacks brow, and often greyish on breast, with spots on throat and breast. Juv. more olive above and dingier below with vaguer undertail-covert tips.

145.7. Long-billed Bush-warbler
Bradypterus major

15cm   p.485

Rather large with long thin bill; olive-brown with short pale brow and pale buff undertail-coverts. Black spots on whitish throat and breast usually form necklace or nearly a gorget, but may be absent, and varies from bright buff to dull whitish on flanks. Juv. is pale yellowish below with mottled malar and breast.

145.8. Sri Lanka Bush-warbler
Elaphrornis palliseri

16cm   p.487

Large, dark and heavy-bodied (suggesting a small babbler); dusky brown with greyish face and rufous patch on lower throat. Juv. plain dark olive above, mottled yellowish-brown below.

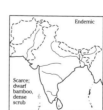

PLATE 146. BIG OR STRIKING REED-WARBLERS (I. Lewington)

Reed-warblers *Acrocephalus* (continued on Plate 147)

Drab brownish warblers of grass and scrub, often skulking and difficult to identify owing to lack of clear field marks. Note details of head pattern and bill, tone of upperparts, primary projection; voice useful. Species shown here are more strikingly marked, or large ('great' reed-warblers).

146.1. Black-browed Reed-warbler
Acrocephalus bistrigiceps

13cm   p.490
Hypothetical; N. Small with bold black lateral crown-stripe (somewhat obscured when fresh) above very broad buffy brow, and unstreaked upperparts.

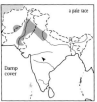

146.2. Moustached Warbler
Acrocephalus melanopogon

13cm   p.489
Small with blackish-brown crown, very bold white brow, dark brown cheek and vaguely streaked rufescent-brown mantle and rufescent rump; white below with bright cinnamon flanks and vent. Lower mandible mostly dark.

146.3. Sedge Warbler
Acrocephalus schoenobaenus   13cm   p.489

Hypothetical; Ladakh. Paler above than Moustached, with streaked coronal stripe, paler yellow-buff underparts and mostly pale lower mandible.

146.4. Oriental Reed-warbler
Acrocephalus orientalis

19cm   p.492
From Indian by slightly more bulbous bill with less dark lower mandible; fine streaks on throat and breast diagnostic but not always present.

146.5. Great Reed-warbler
Acrocephalus arundinaceus

19cm   p.491
Vagrant; NW. Large, with very long primary projection. More rufous than Indian; slightly shorter, thicker bill has just basal third of lower mandible pale. More rufous above than Oriental, without fine breast streaks or yellowish underparts.

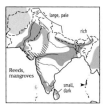

146.6. Indian Reed-warbler
Acrocephalus [stentoreus] brunnescens

19cm   p.492
Only large reed-warbler in most of region; from vagrant Oriental by slightly finer bill (looking almost curved), with more dark on distal lower mandible. No rufous tones above of vagrant Great.

146.7. Thick-billed Warbler
Acrocephalus aedon

20cm   p.493
Large with diagnostic swollen, largely pale flesh bill and lack of dark eyeline on pale face; note rufescent crown (somewhat crested), wings, rump and tail.

PLATE 147. PLAIN REED-WARBLERS AND *HIPPOLAIS* WARBLERS (I. Lewington)

Reed-warblers *Acrocephalus* (continued from Plate 146)

147.1. Large-billed Reed-warbler
Acrocephalus orinus

14cm   p.493
Known from one specimen; russet above with very large bill and entirely pale lower mandible. See text; darker and more richly coloured than Blyth's.

147.4. European Reed-warbler
Acrocephalus scirpaceus

14cm   p.491
Paler above than Blyth's with a shorter weak buff supercilium; longer primary projection than all similar congeners. When worn, pale brown above and nearly white below; more rufescent when fresh.

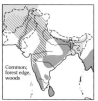

147.2. Blyth's Reed-warbler
Acrocephalus dumetorum

14cm   p.491
Small and olivaceous with longish bill and mostly pale lower mandible, flattish forehead and indistinct brow broadest on lores; below, whitish throat and very long undertail-coverts (pale brown with paler tips). Longer primary projection than similar species except European; less buffy below than Paddyfield and Blunt-winged.

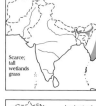

147.5. Blunt-winged Reed-warbler
Acrocephalus concinens stevensi

14cm   p.490

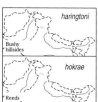

147.6. *Acrocephalus concinens hokrae, haringtoni*

Darker above than both Blyth's and Paddyfield, and more olive above and buffier below than Blyth's; brow is shorter than Paddyfield. Short primary projection, and longish graduated tail narrowest at base. Races differ in soft-part colours and habitat preference; see text.

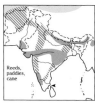

147.3. Paddyfield Warbler
Acrocephalus agricola

14cm   p.490
Note the slim shortish bill with half-dark lower mandible, rufescent rump, long whitish brow with diffuse dark upper margin and short primary projection. In breeder, has dark crown and pale nape; rufescent above when fresh, and orange-buff below with white throat and belly.

Warblers *Hippolais*

Paler than reed-warblers and rather featureless, with pale lores and mostly orange bills; most have peaked crown. Note bill and leg colour, primary projection.

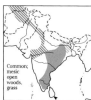

147.7. Booted Warbler
Hippolais caligata

12cm   p.495
Small. Usually shorter-billed, rounder-headed and shorter-tailed than Sykes's, with dark tip to lower mandible and slightly darker legs; not always separable, especially when worn. From Siberian Chiffchaff by mostly pale bill and paler legs; see Paddyfield.

147.9. Eastern Olivaceous Warbler
Hippolais pallida

12.5cm   p.495
Hypothetical; NW. From smaller Sykes's by slightly longer bill with paler lower mandible and longer primary projection. More monochromatic above than larger Upcher's (wings and tail not darker) with shorter primary projection and shorter tail.

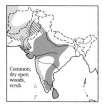

147.8. Sykes's Warbler
Hippolais rama

12cm   p.494
Like Booted but with slightly darker legs and no dark tip to lower mandible or longer bill; flatter forehead and longer, more graduated tail (extending farther past tail-coverts). Paler than Blyth's Reed with a narrower brow and shorter undertail-coverts.

147.10. Upcher's Warbler
Hippolais languida

14cm   p.495
Rather large and long-billed with pale lower mandible; grey above with darker wings and tail, latter longish with broad white tips on underside of outer feathers.

PLATE 148. TIT-WARBLERS, GOLDCREST, TAILORBIRDS AND TESIAS (J. Anderton)

Tit-warblers *Leptopoecile*

Tiny, dimorphic, distinctive tit-like birds with unusual blue, purple and chestnut coloration and red eyes. Note crest, white in tail, eyebrow.

148.1. Crested Tit-warbler
Leptopoecile elegans

10cm p.520
Possible; N Arunachal. Tit-like, with white crest, blue wings and tail, lavender flanks and white outertail. Male is blue above with chestnut head, female chestnut with blackish nape.

148.2. White-browed Tit-warbler
Leptopoecile sophiae

10cm p.521
Uncrested rufous crown, broad pale brow, purplish flanks and vent, and bluish rump and tail. Paler in Kashmir: male is purple-and-blue below with white belly; female has whitish throat and belly. Much darker in C Himalayas: male with chestnut underparts, blue-grey in female.

Tailorbirds *Orthotomus*

Long-billed, olive above with rufous on crown (lacking in juvs.); tail often cocked. Note tail length, head pattern, and vent colour.

148.3. Common Tailorbird
Orthotomus sutorius

13cm p.477
Rather plain olive above and white below with peach flanks and longish narrow tail; rufous cap rather dull. Breeding male with variable tail-spikes. Whitish brow on continent, lacking in Sri Lanka. Juv. (not illustrated) has less rufous on crown.

148.4. Black-necked Tailorbird
Orthotomus atrogularis

13cm p.477
From Common by yellowish flanks, vent and undertail, brighter green upperparts, and usually greyer breast. Male has grey face, streaky black bib, no pale brow. Female may have grey streaks on breast and slight pale brow. Juv. often has little yellow below but has yellowish undertail.

148.5. Mountain Tailorbird
Orthotomus cuculatus

12cm p.476
Bright olive with long white brow, rufous forecrown, silvery cheek and nape with white throat and breast, bright yellow belly, and mostly white outertail. Broad-billed Warbler similar; see Pl. 149. Juv. has yellowish brow and yellow belly and vent; see much shorter-billed Yellow-bellied Warbler (Pl. 149).

Goldcrest *Regulus*

148.6. Goldcrest
Regulus regulus himalayensis group

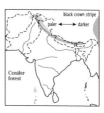

8cm p.521
Tiny; olive-grey with dark bill, black lateral crown-stripe, and white eye-ring, wing-bar and tertial-spots; male has orange crown, yellow in female. Juv. is plain-headed. Lacks dark eyeline of leaf-warblers (Pl. 152); wing-bars more prominent than in non-breeding Fire-capped Tit (Pl. 156).

148.7. *Regulus regulus tristis*

Not illustrated. Much paler and greyer above than Himalayan races, with weak coronal stripes; male has yellow crown-patch, very small in female.

Tesias *Tesia*

Pot-bellied, long-legged and nearly tailless terrestrial warblers, skulking but noisy. Note crown colour.

148.8. Slaty-bellied Tesia
Tesia olivea

9cm p.479
Dark slaty below with brilliant green-gold crown and vivid orange lower mandible; darker overall than Yellow-browed. Juv. (not illustrated) is duller, with all-pale lower mandible.

148.9. Yellow-browed Tesia
Tesia cyaniventer

9cm p.480
From Slaty-bellied by dusky crown above yellowish brow and bolder dark eye-stripe, duller lower mandible and paler grey underparts (whitish at centre of throat and belly). Juv. muddier olive, but still shows pale brow; lower mandible pale with dark tip.

148.10. Chestnut-headed Tesia
Tesia castaneocoronata

8cm p.479
Dark olive with chestnut cap and yellow bib; note white postorbital spot and tiny bill. Juv. is paler and duller, with rufescent cheek, breast and flanks on dilute yellow underparts; told by tiny bill and white postocular spot.

PLATE 149. BRIGHT FOREST WARBLERS AND CANARY-FLYCATCHER (J. Anderton)

Warblers *Tickellia, Abroscopus, Seicercus*

Tiny to mid-sized warblers, often found in large mixed feeding flocks. All are olive with yellow on underparts and white in outertail feathers. Note wing and facial pattern details (eye-ring, brow, etc.), colour of rump.

149.1. Broad-billed Warbler
Tickellia hodgsoni

10cm   p.513
Tiny, with rufous crown, short whitish brow, grey face and breast, and yellow belly; no wing-bars or rump-patch. Longer-billed, longer-browed Mountain Tailorbird (Pl. 148) has dark eyeline.

149.2. Black-faced Warbler
Abroscopus schisticeps

9cm   p.514
Tiny with pink bill and legs; broad yellow brow and throat, with slaty crown and black mask. See Yellow-throated Fulvetta (Pl. 135) and Yellow-bellied Fantail (Pl. 110).

149.3. Rufous-faced Warbler
Abroscopus albogularis

8cm   p.513
Tiny, with diagnostic rufous face-patch below blackish crown-stripe; blackish patch on white throat, and pale yellow rump, breast and vent with white belly. No wing-bars.

149.4. Yellow-bellied Warbler
Abroscopus superciliaris

9cm   p.514
Tiny with heavy dark bill; note grey crown and black eyeline with whitish brow and throat.

149.5. Chestnut-crowned Warbler
Seicercus castaniceps

10cm   p.513
Tiny with black-striped rufous cap and silvery face and breast with white eye-ring, white wing-bars and tail-spots, and yellow rump.

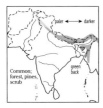

149.6. Grey-hooded Warbler
Phylloscopus xanthoschistos

10cm   p.505
Small with grey crown and mantle with silvery brow between blackish crown-stripe and eyeline; entirely yellow below (vs white throat in Yellow-bellied). No eye-ring or wing-bars.

149.7. Grey-cheeked Warbler
Seicercus poliogenys

10cm   p.515
Dark grey crown and cheek with obscure crown-stripe, broad broken white eye-ring and whitish chin, and yellowish wing-bar. See White-spectacled.

149.8. White-spectacled Warbler
Seicercus affinis

10cm   p.511
From Grey-cheeked by paler crown with clearer black lateral stripe, olive cheek and yellow chin; also yellowish spot on lores, weak wing-bar, and paler lower mandible and legs.

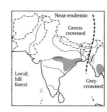

149.9. Green-crowned Warbler
Seicercus burkii

10cm   p.510
Most like Whistler's, but olive crown has bolder black lateral crown-stripe; note narrow yellow eye-ring broken at rear, and wing-bar usually absent.

149.10. Grey-crowned Warbler
Seicercus tephrocephalus   10cm   p.510

Map above. Mostly grey crown and olive cheek with narrow but complete yellow eye-ring; has darker lores and better-defined lateral crown-stripe than Whistler's, and no wing-bar. Outertail feathers half-white.

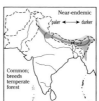

149.11. Whistler's Warbler
Seicercus whistleri

10cm   p.511
From Green-crowned and Grey-crowned by broader (complete) yellow eye-ring, weaker lateral crown-stripe, diffuse yellowish lores and bolder pale wing-bar; note smaller bill with entirely pale lower mandible and orange tint on breast.

Canary-flycatcher *Culicicapa*

An aberrant, distinctive species, somewhat resembling some *Seicercus*.

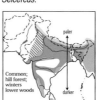

149.12. Grey-headed Canary-flycatcher
Culicicapa ceylonensis

9cm   p.387
Tiny and bright olive above with grey hood lacking eye-ring or crown-stripes, and yellow below. Superficially like *Seicercus*, none of which is grey-throated, and note upright posture.

PLATE 150. BROWN LEAF-WARBLERS (I. Lewington)

Leaf-warblers *Phylloscopus* (continued on Plates 151–152)

Highly active warblers, many given to foraging in outer reaches of foliage. A speciose group, and often very difficult to identify; species on this plate are the plainest, lacking prominent field marks such as wing-bars, tertial-spots, rump-patches or coronal stripes. Note bill and leg colour, details of supercilia and cheeks.

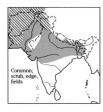

Common; scrub, edge, fields

150.1. Siberian Chiffchaff
Phylloscopus [*collybita*] *tristis*

10cm p.496

Brown and buffy with black bill and legs. In extreme NW, see text for possible Willow Warbler *P. trochilus* (p. 497) and Common Chiffchaff *P. collybita* (p. 497); existing regional records of latter rejected (Appendix 2).

Breeds montane scrub; winters riverine scrub

150.2. Mountain Chiffchaff
Phylloscopus sindianus

10cm p.497

Like Siberian but ruddier (fresh) or sandier (worn) above with no trace of olive in hindquarters, whiter brow wide in front of eye and ending shortly behind it, less marked eyeline, and brighter underparts with whiter throat. Very similar to Plain Leaf-warbler.

Breeds in junipers; winters tamarisks

150.3. Plain Leaf-warbler
Phylloscopus neglectus

9.5cm p.497

Tiny and very pale with short tail and small dark bill; plain grey above, rather faint whitish brow, and whitish below.

a paler race

Common in E; wetlands grass, thickets

150.4. Dusky Warbler
Phylloscopus f. fuscatus

11cm p.498

Plain and dark with fine dark bill and slightly rounded tail; brown above, usually greyish on breast and buffy on vent. Variable, but narrow whitish brow clear in front of eye, often less distinct behind (reverse of Radde's). Siberian Chiffchaff has darker legs, weaker brow.

a very dark race

Scarce; grass, thickets

150.5. *Phylloscopus fuscatus weigoldi*

NE wintering race *weigoldi* darker overall than nominate, approaching Smoky.

150.6. Radde's Warbler
Phylloscopus schwarzi 12.5cm p.499

Hypothetical; N, Bangladesh. Brownish, with shortish, heavy, mostly pale bill, heavy, very pale legs, and buffy breast-band. Pale brow (with dark line above) usually indistinct buff in front of eye (well marked in Dusky). In NE, see text for possible Yellow-streaked *P. armandii* (p.499).

Near-endemic

Scarce; alpine scrub, reeds

150.7. Smoky Leaf-warbler
Phylloscopus fuliginventer

10cm p.498

Very dark olive-brown above with a dull, pale brow, longish, mostly dark bill, and dull dark brown below with dark undertail-coverts; note the broad, rather short squared dark tail. Dusky has indistinct brow and more rounded tail; Tickell's is usually paler and has mostly pale lower mandible.

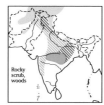

Rocky scrub, woods

150.8. Sulphur-bellied Warbler
Phylloscopus griseolus

11.5cm p.499

Plain, with rather heavy darkish bill; drab brown (no olive) above with strong brow yellowish in front of eye; dull greyish-yellow below. More yellow below than Dusky, less than Tickell's.

Breeds alpine scrub, winters woods

150.9. Tickell's Leaf-warbler
Phylloscopus affinis

11cm p.498

Plain with bright yellow brow over bold blackish eyeline, yellow underparts, and rather long bill with yellowish lower mandible; greener above than Sulphur-bellied, paler above (usually) than Smoky, and shorter-tailed than Aberrant Bush-warbler (Pl. 144). See Buff-throated.

Scrub

150.10. Buff-throated Leaf-warbler
Phylloscopus subaffinis

10.5cm p.499

Vagrant; S Assam hills. Very like Tickell's but lower mandible mostly dark, facial pattern usually less distinct, and brow, throat, breast and flanks buffier; see also Radde's Warbler and much longer-tailed Aberrant Bush-warbler (Pl. 144).

PLATE 151. GREEN LEAF-WARBLERS (I. Lewington)

Leaf-warblers *Phylloscopus* (also on Plates 150, 152)

These species are greener than those on Pl. 150, but mostly plainer than those on Pl. 152, lacking tertial spots, strong head pattern, and rump-patches.

151.1. Tytler's Leaf-warbler
Phylloscopus tytleri

10cm p.504
Small and drab with very long whitish eyebrow over strong eye-stripe, fine long dark bill and no wing-bar: olive above and white below (fresh) or greyish above, dingy below (worn).

151.2. Greenish Warbler
Phylloscopus trochiloides viridanus

11cm p.502
Mid-sized with dull olive upperparts, pale lower mandible, and strong whitish eyeline and single wing-bar; most similar to worn Bright-green (see text). Paler than nominate Greenish, with dull whitish unmottled underparts, olive-tinged flanks and yellowish belly, and weaker face pattern.

151.3. *Phylloscopus t. trochiloides*

Like *viridanus* race of Greenish, but with mostly dark bill, stronger facial pattern, darker upperparts and dingy, mottled breast.

151.4. Two-barred Warbler
Phylloscopus plumbeitarsus

p.503
Like nominate Greenish but (when fresh) has longer, broader, more clear-cut bars on greater and median coverts. Usually slightly greener above, with whiter underparts, and mostly pale lower mandible. In worn plumage, not safely identifiable.

151.5. Bright-green Warbler
Phylloscopus nitidus

12cm p.503
Brighter green above (usually) than Greenish with stouter bill; see text, especially for *viridanus* Greenish. Whitish wing-bar, wide yellowish supercilium, and yellow-tinged throat and breast in fresh plumage; greyer above and whiter below when worn.

151.6. Arctic Warbler
Phylloscopus borealis

12cm p.502
Vagrant; Andamans. Drab, heavy bill, single to two small wing-bars, long wing; long pale brow does not reach bill. Very like Greenish and Large-billed; see text.

151.7. Large-billed Leaf-warbler
Phylloscopus magnirostris

12.5cm p.503
Large, with large dark bill and single wing-bar, long yellowish brow over broad eyeline and finely mottled yellowish cheek. Very like Greenish and Arctic; see text.

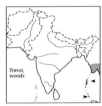

151.8. Pale-legged Leaf-warbler
Phylloscopus tenellipes

11cm p.503
Smallish with fleshy legs and one to two narrow white or buffy wing-bars, and bright white brow over broad eye-stripe. Olive above with brownish rump and tail, grey or brown crown, and white below with mottled greyish throat and breast and fulvous-tinged flanks.

151.9. Eastern Crowned Warbler
Phylloscopus coronatus 11.5cm p.504

Hypothetical; NE. Fairly large, with pale coronal stripe on grey crown, one narrow wing-bar (*vs* two in Blyth's Leaf), and white underparts with (usually) yellowish vent (*vs* greyish throat and whitish vent in Western Crowned).

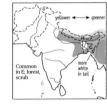

151.10. Blyth's Leaf-warbler
Phylloscopus reguloides

11cm p.505
Bright green with two broad yellowish wing-bars, broad white outertail-tips and strong yellowish brow; yellowish below. Shorter-billed than Western Crowned, with two pale wingbars. In NE, see text for possible White-tailed *P. davisoni* (p. 505).

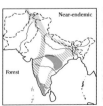

151.11. Western Crowned Warbler
Phylloscopus occipitalis

11.5cm p.504
From very similar Blyth's by narrower wing-bars (upper may be lacking). Paler and greyer above than Eastern Crowned, with weaker crown-stripe and eye-stripe, yellower brow, and more white on tail-tip; whitish below with fine yellow streaks, greyish throat and breast, and almost no yellow on vent.

PLATE 152. LEAF-WARBLERS WITH TERTIAL SPOTS (I. Lewington)

Leaf-warblers *Phylloscopus* (continued)

These species (except distinctive Yellow-vented) told by white or whitish trim on outer web of tertials; most also have bold head patterns with coronal stripe, pale brow and eye-stripe. Note also presence of rump-patch and white in tail. See text for juveniles.

152.1. Yellow-browed Leaf-warbler
Phylloscopus inornatus

10cm   p.501

Fairly bright green with two bold yellowish wing-bars, large white tertial tips, green crown (sometimes with paler coronal stripe), and yellowish brow; paler and clearer yellow below than Mandelli's and Hume's, with pale legs and lower mandible and more prominent tertial spots. Lacks pale rump-patch of Lemon-rumped.

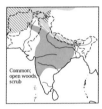

152.2. Hume's Leaf-warbler
Phylloscopus humei

10cm   p.502

Drabber than Yellow-browed, with dark bill and legs. When fresh, grey-green above with brighter green rump, two buff wing-bars and pale buffish eyebrow over strong eyeline, and yellow-buff wash below. Very similar Mandelli's is yellower below with greyer throat.

152.3. Mandelli's Leaf-warbler
Phylloscopus [humei] mandellii

10cm   p.501

Very like Yellow-browed, but with mostly dark lower mandible and darker legs; also bolder eye-stripe, and in fresh plumage, buffy wing-bars. Even more similar to Hume's, but note greyish throat on dull yellow underparts; see text.

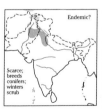

152.4. Brooks's Leaf-warbler
Phylloscopus subviridis

10cm   p.501

Olive above with paler yellowish rump (not a clear patch), double whitish wing-bars, dark olive (not black) lateral crown-stripe and yellowish coronal stripe and brow; always some yellow at base of lower mandible. Very like Yellow-browed and Hume's, but more yellowish overall.

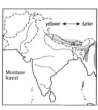

152.5. Orange-barred Leaf-warbler
Phylloscopus pulcher

10cm   p.500

Tiny, with longish fine black bill, bright buff-orange wing-bars, yellow rump and white tail-spots; note smudgy breast, lacking in Grey-faced, which has yellow wing-bars.

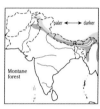

152.6. Lemon-rumped Leaf-warbler
Phylloscopus chloronotus

9.5cm   p.500

Tiny, with no white in short tail; olive crown with yellow coronal stripe and brow, double wing-bars and rump; underparts yellowish.

152.7. Grey-faced Leaf-warbler
Phylloscopus maculipennis

9cm   p.500

Tiny, with white inner webs on short tail and short dark bill; strong head pattern, bold yellow double wing-bar and rump-patch. Note silvery throat and yellow belly. Lemon-rumped lacks white in tail.

152.8. Chinese Leaf-warbler
Phylloscopus yunnanensis

10cm   p.500

See text. Very like Lemon-rumped, with pale lower mandible of longer bill, paler legs, and no dark patch at base of secondaries. Lacks white in tail of Orange-barred and Grey-faced.

152.9. Yellow-vented Warbler
Phylloscopus cantator

9cm   p.505

Very distinctive; yellow head with bold black crown-stripes and eyeline, and yellow breast and vent; no white in tail. Bright olive above with narrow yellow wing-bar.

PLATE 153. *SYLVIA* (I. Lewington)

Warblers *Sylvia*

Rather stocky, plain warblers with thick bills and whitish outertail feathers. All are whitish below, and (except Barred) lack wing-bars. Note head pattern, and eye, bill and leg colour.

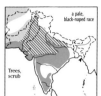

153.1. Eastern Orphean Warbler
Sylvia crassirostris

15cm   p.518

Very large and heavy-billed with blackish cheek and tail, and white underparts. Male is greyish above with black cap and nape (pale in smaller male Ménétries's), and pale eye but no eye-ring. Female brownish with grey crown, dark mask and eye. First-winter browner, with dark eye. In extreme NW, see text for possible Blackcap *S. atricapilla* (p. 516).

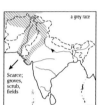

153.2. Common Whitethroat
Sylvia communis

12cm   p.518

Rather large and pale brown with rufous wing, white throat and broken eye-ring, orange eye, pale flesh-toned legs and dark-tipped yellowish bill. Male has grey cap.

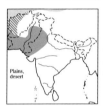

153.3. Asian Desert Warbler
Sylvia nana

11cm   p.520

Tiny; sandy-brown above with pale rufous rump and tail, whitish outertail feathers, and yellow eye, bill and legs.

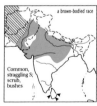

153.4. Lesser Whitethroat
Sylvia curruca halimodendri

12cm   p.517

Mid-sized; warm brown above with darker brown tail, dark cheek-patch, and grey crown. Bill is shorter than Hume's with more pale at base; longer and thicker than in 'Desert' Whitethroat.

153.5. 'Desert' Whitethroat
Sylvia [*curruca*] *minula*

<12cm   p.517

Smaller and paler than Lesser and Hume's, with a partly pale bill; brownish above with (only) slightly darker tail and pale grey crown with diffuse pale brow above dark mask.

153.6. Hume's Whitethroat
Sylvia althaea

>12cm   p.518

Mid-sized with black legs and bill; uniformly darker grey above than Lesser and 'Desert' Whitethroats.

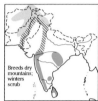

153.7. Ménétries's Warbler
Sylvia mystacea

12cm   p.520

Small; blackish tail with white outer feathers, pale lower mandible and reddish legs. Male is grey above with black crown and cheek, red eye and orange eye-ring; much smaller than male Eastern Orphean, with pale nape. Female and first-winter are sandy-brown above with broken eye-ring; lack masked appearance of whitethroats, rufous tail of Asian Desert.

153.8. Garden Warbler
Sylvia borin   15cm   p.516

Hypothetical; Ladakh. Large, drab olive-brown with short thick greyish bill, dark legs and eye; narrow whitish eye-ring, slight brow and greyish post-auricular patch. Most reed-warblers (Pl. 146, 147) are shorter-winged with longer, paler bills and different overall shape.

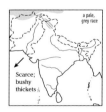

153.9. Barred Warbler
Sylvia nisoria

16cm   p.517

Large with thick bill, yellow eyes; grey above with black-and-white scales and wing-bars.

PLATE 154. GREY TITS (J. Anderton)

Crested tits Parus

Small and crested, most black-headed with prominent white cheek-patches and some rufous (or buff) on underparts. Note colour of flanks, belly and vent, size of black bib, and wing pattern.

'Typical' tits Parus

All the following are uncrested with black heads and ventral stripes and prominent white cheek-patches. Note back colour, wing and cheek pattern.

154.1. Rufous-vented Tit
Parus rubidiventris

10cm   p.525
Mid-sized with black bib on throat only; lacks wing-bars. Belly and vent rufous in W Himalayas. Mostly grey with rufous-tinted vent in E, without rufous flank-patches.

154.2. Rufous-naped Tit
Parus rufonuchalis

13cm   p.525
Bigger than similar species; grey below with black bib extending to upper belly and only vent and flank-patch rufous.

154.3. Spot-winged Tit
Parus melanolophus

11cm   p.526
Small with a double wing-bar of white spots; flanks rufous, and belly grey.

154.4. Coal Tit
Parus ater

10cm   p.526
Only white-bellied crested tit in region; wing-pattern like Spot-winged with which it hybridises in W Nepal.

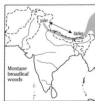

154.5. Grey-crested Tit
Parus dichrous

12cm   p.526
Long-crested and rather monochromatic; grey above and buff below with pale half-collar.

154.6. White-naped Tit
Parus nuchalis

13cm   p.528
Glossy black above with 'open' white cheek-patch, and large white wing-panel; no grey.

154.7. Great Tit
Parus major

13cm   p.527
White cheek-patch broadly enclosed by black, and single white wing-bar on blue-grey wing; back ranges from pale grey (NW; very like Turkestan) to almost slaty (S and Sri Lanka) to yellow-tinted in Tibet (possible in NE Himalayas).

154.8. Black-bibbed Tit
Parus hypermelaenus

11cm   p.525
Possible; S Assam hills. Brownish above and uncrested, no ventral stripe. In N Arunachal, possible Willow *P. montanus weigoldicus* (p. 525) is larger, browner, with more cinnamon flanks.

154.9. Turkestan Tit
Parus bokharensis

15cm   p.528
Similar to NW race of Great; note larger, more triangular white cheek-patch with narrower (and sometimes broken) rear black border, and blue-grey inner webs of tertials. See text.

154.10. Green-backed Tit
Parus monticolus

13cm   p.527
Unmarked olive above with two white wing-bars and yellow underparts; combination of white cheek and yellow belly diagnostic, but beware possible Tibetan form of Great, and juvs. of other tits.

PLATE 155. LONG-TAILED TITS AND COLOURFUL TITS (J. Anderton)

Long-tailed tits *Aegithalos*

Tiny, long-tailed and extremely active, restricted to the Himalayas and recalling some parrotbills (Pl. 140) and fulvettas (Pl. 135). Most adults are pale-eyed; juvs dark-eyed and usually lack prominent throat/breast-marks. Note head pattern.

155.1. Red-headed Tit
Aegithalos concinnus

10cm p.522
Rufous crown, black mask and throat-spot; breastband and flanks chestnut in S Assam hills. Juv. lacks throat-spot.

155.2. Rufous-fronted Tit
Aegithalos iouschistos

10cm p.522
Fulvous with broad black mask and silvery throat. Juv. is similar but paler buff.

155.3. 'Black-browed' Tit
Aegithalos [iouschistos] bonvaloti

10cm p.522
Form *bonvaloti* possible in Arunachal; like Rufous-fronted but much paler, with white crown and throat. See text for distinctive race *sharpei*, possible in e.g. Lushai Hills.

155.4. White-cheeked Tit
Aegithalos leucogenys

10cm p.522
Brown crown, narrow black mask, white cheek-patch, and full black bib. Juv. lacks bib.

155.5. White-throated Tit
Aegithalos niveogularis

10cm p.522
White forecrown, throat and half-collar; broad dark mask and dark eye. Juv. duller.

Miscellaneous tits *Pseudopodoces, Parus*

155.6. Ground-tit
Pseudopodoces humilis

20cm p.535
Starling-sized and very pale, with thin decurved black bill and whitish-sided black tail; formerly considered a corvid but now known to belong in tits Paridae. Suggests a wheatear (Pl. 121). See also Pl. 178.

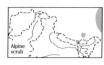

155.7. White-browed Tit
Parus superciliosus

14cm p.527
Possible; NE. Buff-brown with black crown, eyeline and throat, and long white brow.

155.8. Azure Tit
Parus cyanus

13cm p.529
Very like Yellow-breasted but breast white.

155.9. Yellow-breasted Tit
Parus flavipectus

13cm p.530
Whitish with blue wings and tail, lemon-yellow breast, and broad white wing-bar. Juv. is olive-tinted on mantle and washed yellow below; may not be separable from juv. Azure. See text (p. 530) for possible Blue Tit *P. caeruleus*.

Yellow tits and Sultan Tit *Parus, Melanochlora*

Yellow tits are olive-backed with prominent yellow-tipped dark crests, pale wingbars on grey wings, and dark ventral stripes. Note facial and mantle patterns. Sultan Tit much larger but rather similar.

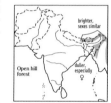

155.10. Black-spotted Yellow Tit
Parus spilonotus

14cm p.529
Very like Black-lored but forehead and lores yellow, eyeline short, and mantle streaked.

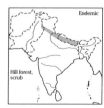

155.11. Black-lored Yellow Tit
Parus xanthogenys

14cm p.528
Sexes alike. Note black forehead, lores and curved border to cheek, and unmarked mantle. Very like Indian, but wing-bars are yellow and black bib and crest are glossier.

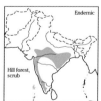

155.12. Indian Yellow Tit
Parus aplonotus

14cm p.529
Like Black-lored but wing-bars white; may have fine yellow brow over lores. Female duller than male with olive rather than black ventral stripe; may almost lack yellow, especially in S.

155.13. Sultan Tit
Melanochlora sultanea

20cm p.530
Large and unmistakable, with long, loose yellow crest, dark upperparts (black in male, dark olive in female) and yellow belly.

PLATE 156. CREEPERS AND ABERRANT TIT-LIKE BIRDS (J. Anderton)

Treecreepers *Certhia*

Slim brown inconspicuous Himalayan birds normally seen clinging close to and spiralling up tree-trunks, using tail as prop (see also nuthatches, Pl. 157). Often in feeding flocks. Note face pattern, bill length, colour of underparts and barring on tail.

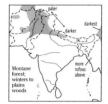

156.1. Bar-tailed Treecreeper
Certhia himalayana

12cm p.541
Only treecreeper with a barred tail; note very long, curved bill. Lacks rufous or buff below.

156.2. Eurasian Treecreeper
Certhia familiaris hodgsoni

12cm p.541
From Bar-tailed by shorter bill and unbarred tail with short spikes, and usually more rufescent flanks and rump. NW *hodgsoni* pale, longer-billed.

156.3. *Certhia familiaris mandellii*

Darker above with more rufous rump, and shorter bill in C and E Himalayan *mandellii*.

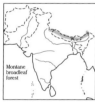

156.4. Rusty-flanked Treecreeper
Certhia nipalensis

12cm p.542
Short straight bill; note rufous flanks, nearly black crown and mantle, and broad brow enclosing dark cheek.

156.5. Brown-throated Treecreeper
Certhia discolor

12cm p.543
Longish curved bill and rufescent tail; throat dull brown in Himalayas, cinnamon in juv. and S Manipur and Mizoram.

Creepers *Tichodroma, Salpornis*

156.6. Wallcreeper
Tichodroma muraria

17cm p.540
Unique; slate-grey with long curved bill, magenta wing, and shortish squared tail. Throat is black in breeding, white in non-breeding. In flight, wings very rounded with bold white spots in primaries.

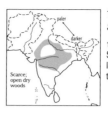

156.7. Spotted Creeper
Salpornis spilonotus

13cm p.543
Stout with long curved bill; treecreeper-like but with short squared tail and spotted underparts.

Tits and Penduline tits *Sylviparus, Remiz, Cephalopyrus*

Remiz species are tiny, short-tailed tits with pointed bills and usually some chestnut above. Taxonomy complex.

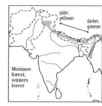

156.8. Yellow-browed Tit
Sylviparus modestus

9cm p.535
Tiny and plain greyish-olive, round-crested and stubby-billed, with weak pale brow and wing-bar. More monochromatic than Fire-capped Tit; see Green Shrike-babbler (Pl. 137).

156.9. 'Black-headed' Eurasian Penduline Tit
Remiz pendulinus macronyx

10cm p.523
Longer-billed than White-crowned; fulvous with black hood, chestnut mantle and rufous flanks.

156.10. Eurasian Penduline Tit
Remiz pendulinus caspius

Map above. Bigger-billed than White-crowned, male with chestnut crown; female (not shown) has darker crown than White-crowned, and is browner above.

156.11. White-crowned Penduline Tit
Remiz coronatus

10cm p.524
Pale with buffy rump, tiny bill. Male has white crown, black mask broadening at rear, and chestnut band on mantle. Female has weak mask, little chestnut; juv. very plain, as with juv. congeners.

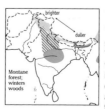

156.12. Fire-capped Tit
Cephalopyrus flammiceps

9cm p.524
Tiny; olive with conical bill, pale wingbars, and short notched tail. Breeding male distinctive; breeding female is duller with whitish belly. Non-breeders and juv. plain, greyer, with yellow-trimmed wings.

PLATE 157. NUTHATCHES (J. Anderton)

Nuthatches *Sitta*

Small insectivores specialised in foraging on tree bark, climbing (often upside-down) without use of their short squared tails (see also treecreepers, Pl. 156); one regional species is almost entirely terrestrial. Most are grey above and white or buff below with broad black eye-stripes; sexes essentially similar. Note colour and pattern of face, undertail-coverts.

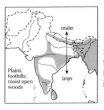

157.1. Velvet-fronted Nuthatch
Sitta frontalis
10cm   p.538
Small; blue above with red bill, black forehead and yellow eye and eye-ring. Male has a narrow black eyeline, missing in female.

157.7. White-cheeked Nuthatch
Sitta leucopsis
12cm   p.537
White face and underparts with chestnut vent, and black crown but no eye-stripe.

157.2. Kashmir Nuthatch
Sitta cashmirensis
12cm   p.536
Rather large, large-billed, and rather pale; rufous-buff below including unmarked vent.

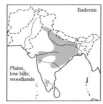

157.8. Indian Nuthatch
Sitta castanea
12cm   p.536
Chestnut below with white cheek-patch; from similar Chestnut-bellied by shorter, slimmer bill, slightly paler crown, and grey undertail-coverts with chestnut trim. Male is slightly darker below than male Chestnut-bellied, with more white on chin, less on cheek.

157.3. Chestnut-vented Nuthatch
Sitta nagaensis
12cm   p.535
In S Assam hills, whitish to pale grey below with chestnut rear flanks and vent, latter with white spots. In Arunachal, possible *montium* has somewhat buffier breast.

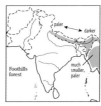

157.9. Chestnut-bellied Nuthatch
Sitta cinnamoventris
12cm   p.536
Very like Indian but with a longer, thicker bill and whitish undertail-coverts edged rufous; crown concolorous with mantle. Male usually has narrower white chin-strap than male Indian.

157.4. White-tailed Nuthatch
Sitta himalayensis
12cm   p.537
Smaller, smaller-billed and usually yellower below than Kashmir; white patch at base of uppertail difficult to see in field.

157.10. Eastern Rock Nuthatch
Sitta tephronota
15cm   p.538
Large and heavy-billed; pale grey above with black eyeline, and whitish below.

157.5. White-browed Nuthatch
Sitta victoriae
12cm   p.537
Possible; extreme SE. Like White-tailed but with white brow and breast, narrow mask. In extreme NE, see text for possible Yunnan *S. yunnanensis* (p. 537).

157.11. Beautiful Nuthatch
Sitta formosa
15cm   p.540
Large; blue with black-and-white streaks above, rufous below with white face.

157.6. Przewalsky's Nuthatch
Sitta przewalskii
12cm   p.537
Possible; N Arunachal. Like White-cheeked but with smaller bill and bright rufous underparts.

Plate sponsored by Mr. Perry Bass.

PLATE 158. FLOWERPECKERS (L. McQueen)

Flowerpeckers *Dicaeum*

Small nectarivorous birds, normally hard to see well owing to size, speed and canopy-dwelling habits. Some are very dull (especially females and juveniles) and can recall some female sunbirds (Pl. 159–160), but are shorter-billed and shorter-tailed. Juveniles have pale bills. Note bill shape and colour, pattern of underparts and rump.

158.1. Pale-billed Flowerpecker
Dicaeum erythrorhynchos

8cm   p.544

Small, drab and greyish, with rather heavy, mostly pale flesh, curved bill. Darker in Sri Lanka, where only small, plain, unstreaked flowerpecker. Nilgiri is darker above with a pale brow and a bluish, dark-tipped bill.

158.2. Nilgiri Flowerpecker
Dicaeum concolor

9cm   p.545

Like allopatric Plain but larger, with broader whitish brow and heavier bluish bill with black tip. Darker above than Pale-billed and Thick-billed.

158.3. Andaman Flowerpecker
Dicaeum virescens

8cm   p.545

Only flowerpecker in range. Bright olive-green with speckled crown, pale face and brow, greyish breast, yellow belly and glossy black tail.

158.4. Plain Flowerpecker
Dicaeum minullum

8cm   p.545

Tiny and drab; fine curved black bill with pale base to lower mandible. Darker and greener above than very similar (but heavier-billed) Pale-billed. From female Fire-breasted by darker crown and paler brow, less buffy underparts. As with congeners, juv. has pale bill (see Pale-billed).

158.5. Thick-billed Flowerpecker
Dicaeum agile

9cm   p.543

Dull greyish with stout stubby bill, reddish eye, diffusely streaked underparts, and white tail-spots. More heavily streaked below in NE with slimmer bill and pale yellowish vent; see juv. Yellow-vented, which has a longer, curved bill.

158.6. Yellow-vented Flowerpecker
Dicaeum chrysorrheum

9cm   p.544

Rather large with curved bill; olive above with red eye, white lores, white underparts with bold black streaks (much stronger than in any other flowerpecker) and yellow vent. Juv. is duller and weakly streaked below; told from NE Thick-billed by bill shape.

158.7. Scarlet-backed Flowerpecker
Dicaeum cruentatum

7cm   p.546

In both sexes, red rump, buff central underparts, and longish curved bill. Male is black above with broad red stripe from forehead to base of tail. Female grey-brown above; see eclipse Fire-tailed Sunbird (Pl. 160), which has longer bill and longer olive tail. Juv. has orange-tinted rump.

158.8. Fire-breasted Flowerpecker
Dicaeum ignipectus

7cm   p.545

Small straight bill and buffy central underparts. Male is blue-black above with bright red breast-patch and black stripe on belly; female is dark olive above with narrow eye-ring.

158.9. Orange-bellied Flowerpecker
Dicaeum trigonostigma

9cm   p.544

Longish heavy bill, blackish tail and yellow-orange rump. Male is slaty above with pale grey throat and orange belly; female is greyish-olive with whitish throat, and orange-yellow belly.

158.10. Yellow-bellied Flowerpecker
Dicaeum melanoxanthum

12cm   p.544

Large and stubby-billed with red eye, white throat and stripe on centre of breast, yellow belly and white outertail-spots. Male is blue-black above and on breast-sides; female browner above and washed-out below.

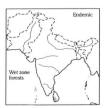

158.11. Legge's Flowerpecker
Dicaeum vincens

9cm   p.544

Unique in range: stout-billed with dark upperparts, white throat, yellow belly and large white tail-tips. Male is blackish above, female brownish-grey. Similar only to larger allopatric Yellow-bellied.

PLATE 159. WHITE-EYES AND SHORT-TAILED SUNBIRDS (L. McQueen)

White-eyes *Zosterops*

Small, active, yellow-and-olive nectarivorous birds with short pointed bills and very conspicuous white eye-rings. Note overall coloration, bill size.

159.1. **Oriental White-eye**
Zosterops palpebrosus

10cm p.551

Small, golden-olive with a very bold white eye-ring, yellowish forehead (varies regionally), yellow throat and vent and silvery to brownish-tinged sides. Some birds in S Western Ghats recall Ceylon.

159.2. **Ceylon White-eye**
Zosterops ceylonensis

11cm p.551

Larger and duskier than Oriental, with longer, slightly curved bill, duller throat, grey flanks, and blackish tail.

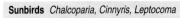

Sunbirds *Chalcoparia, Cinnyris, Leptocoma*

Small and plump with short, squared tail; very active. Males very glossy and colourful in direct light, black in poor light; some adopt eclipse plumage intermediate between female and full male plumage. Note for males colour of belly, crown and bib; for females colour of throat and rump, presence of white in tail.

159.3. **Ruby-cheeked Sunbird**
Chalcoparia singalensis

10cm p.546

Bill short and straight; note rufous throat and yellow belly. Male emerald-green above with bronzy cheek. Female yellow-olive above; juv. is much drabber but has yellow throat, and lacks eye-ring of white-eyes.

159.4. **Olive-backed Sunbird**
Cinnyris jugularis (part)

10cm p.547

Olive above with yellow underparts and large white tail-tips. Male has glossy blue throat; in eclipse, broad glossy stripe from chin to breast like eclipse Purple, but much more olive above. Female has whitish brow. Andamans male lacks shiny forecrown; belly pale yellow.

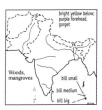

159.5. *Cinnyris jugularis* (Nicobars)

In Nicobars, underparts rich yellow; breeding male has blue-purple forecrown and gorget. Bill shortest on Car Nicobar, intermediate in C Nicobars, and largest on S Nicobars.

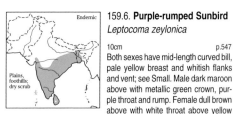

159.6. **Purple-rumped Sunbird**
Leptocoma zeylonica

10cm p.547

Both sexes have mid-length curved bill, pale yellow breast and whitish flanks and vent; see Small. Male dark maroon above with metallic green crown, purple throat and rump. Female dull brown above with white throat above yellow breast; lacks red rump of female Small.

159.7. **Small Sunbird**
Leptocoma minima

8cm p.547

Tiny, with shortish bill. Male paler above than male Purple-rumped with broader dark breast-band and pale yellow vent. Female more olive above than female Purple-rumped with dark red rump and plain underparts; eclipse male similar with brighter olive head and red on mantle.

159.8. **Van Hasselt's Sunbird**
Leptocoma brasiliana

9cm p.547

Tiny, shortish-billed. Male black above with green crown, diagnostic blue scapulars and rump, purple throat, maroon breast and dark vent; lacks yellow below. Female is plain, with narrow pale eye-ring and dark cheek; shorter-tailed than female *Aethopyga* sunbirds.

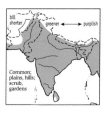

159.9. **Purple Sunbird**
Cinnyris asiaticus

10cm p.548

Smallish; similar in both sexes to Loten's, but bill shorter and less curved. Male black with metallic purplish foreparts, and sometimes a slight maroon breast-band. Female pale brown above with pale brow and dark mask, and variably yellowish below with white outer-tail-tips; eclipse male similar with blue-black ventral stripe, shiny blue shoulder and no tail-spots.

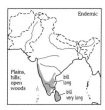

159.10. **Loten's Sunbird**
Cinnyris lotenius

13cm p.548

Bill longer and more curved than in smaller Purple, and tail more rounded. Male greener-glossed than male Purple, with broad maroon breast-band and browner wings and belly. Female darker above, contrasting more sharply than female Purple with greyer underparts. Immature male has broad black stripe from chin to breast and green 'shoulders' and rump; apparently no eclipse.

PLATE 160. LONG-TAILED SUNBIRDS AND SPIDERHUNTERS (L. McQueen)

Sunbirds *Aethopyga*

Both sexes longer-tailed than in *Cinnyris* and *Leptocoma* spp. (Pl. 159); males extremely colourful, with tail-streamers (sometimes missing), but with gloss restricted to head and tail. For males, note colour of mantle, rump and belly, also of glossy bib and crown-patches; in females colour of rump and underparts, white in tail.

160.1. Vigors's Sunbird
Aethopyga vigorsii

male 15cm; female 10cm   p.550
Darker and bigger-billed than Crimson, male with yellow streaks on breast, smaller yellow back-patch and shorter tail; female darker and greyer than female Crimson.

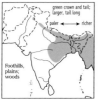

160.2. Crimson Sunbird
Aethopyga siparaja (part)

male 15cm; female 10cm   p.549
Male red with metallic cap and whisker, large yellow back-patch, and olive belly. Female is plain olive with rounded tail (no pale tips), unstreaked belly and no yellow on back. Subadult male like female, but patchily red. On mainland male's cap and tail green.

160.3. *Aethopyga siparaja nicobaricus*

10cm   p.549
In Nicobars, smaller, male with purple cap, rump and tail, no streamers, more orange breast, and olive-grey belly.

160.4. Fire-tailed Sunbird
Aethopyga ignicauda

male 19cm; female 10cm   p.550
Male glowing red-orange above with very long tail; yellow below. Female olive with greyish hood, rather straight bill, yellowish belly and rump, and orange-tinted tail lacking pale tips. First-winter male bigger, with more orange rump and tail than female.

160.5. Mrs Gould's Sunbird
Aethopyga gouldiae

male 15cm; female 10cm   p.548
Both sexes rather short-billed with prominent yellow back-patch. Male red above and yellow below with blue tail. Female olive with yellowish belly, large white tips on outertail; subadult male similar, but patchily red with greyer throat, brighter belly and back, and longish metallic tail. Red-breasted *dabryi* in vagrant to NE.

160.6. Black-breasted Sunbird
Aethopyga saturata

male 15cm; female 10cm   p.549
Both sexes have narrow yellow back-patch and whitish flank tufts. Male looks black with pale belly, maroon mantle, glossy purple moustache and crown, and steely-blue rump and tail. Female like female Mrs Gould's but colder olive overall with indistinct grey tail-tips, no yellow on belly.

160.7. Green-tailed Sunbird
Aethopyga nipalensis

male 15cm; female 10cm   p.549
Male yellow below, olive above with bottle-green head and tail, maroon mantle, and yellow back-patch. Female dull olive with greyish hood, vaguely brighter rump and yellowish belly; has longer brighter olive tail than similar female sunbirds, with large pale tips to outer feathers. In NW, male has little maroon above; in Mizoram, lacks maroon above.

Spiderhunters *Arachnothera*

Sunbird-like but drab, with very long heavy-based bills and strong legs.

160.8. Streaked Spiderhunter
Arachnothera magna

17cm   p.551
Chunky, bright olive and entirely streaked, with very long bill and heavy orange legs.

160.9. Little Spiderhunter
Arachnothera longirostra

14cm   p.550
Smallish and unstreaked, with yellow belly, whitish brow, dark whisker-mark, and bluish legs. Similar female Loten's Sunbird (Pl. 159) is smaller and straighter-billed without a pale brow.

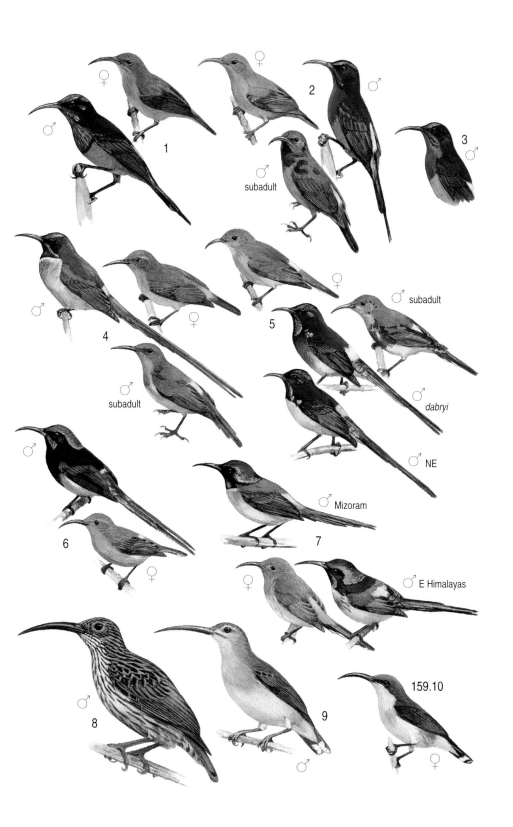

PLATE 161. ACCENTORS (L. McQueen)

Accentors *Prunella*

Slim, fine-billed montane birds, with neatly patterned, earth-toned plumage; superficially pipit-like, but not usually occurring together. Note especially pattern of head and underparts, strength of streaks on mantle.

161.1. Altai Accentor
Prunella himalayana
15cm   p.354
Smaller than similar Alpine with diagnostic rufous streaks on breast; paler above and below, with bolder streaks on mantle. Juv. from juv. Alpine by weaker streaks below, and spotted breast.

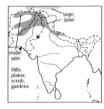

161.5. Black-throated Accentor
Prunella atrogularis
15cm   p.355
Black throat is diagnostic, but partly obscured when fresh. Darker above than Brown and Radde's, with darker border to broad pale brow, heavier streaks on mantle, and brighter buff breast. Juv. has darker throat than similar species.

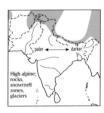

161.2. Alpine Accentor
Prunella collaris
17cm   p.354
Large and grey, with yellow-based bill, scaly throat, and white spots on wing-coverts and tail-tips; rufous streaks on flanks only. Juv. is heavily streaked but has adult's wing and tail pattern.

161.6. Radde's Accentor
Prunella ocularis
16cm   p.355
Darker above than Brown, especially on crown, with heavier streaks on mantle and flanks, fine-spotted malar line, and buff breast-band.

161.3. Robin Accentor
Prunella rubeculoides
17cm   p.354
Plain grey head, rufous breast and nearly unstreaked white belly. Juv. has white belly and streaked brownish breast.

161.7. Brown Accentor
Prunella fulvescens
15cm   p.355
Paler than Black-throated and Radde's, with scarcely streaked upperparts and white throat. Juv. is paler and less streaked than comparable juvs., with pale brow and solid brown cheek, finely streaked breast-band, and nearly unstreaked pale buff throat and belly. Siberian *P. montanella* rejected for region (Appendix 2).

161.4. Rufous-breasted Accentor
Prunella strophiata
15cm   p.354
Rufous brow and breast-band (buffier in W), and blackish cap and cheek. Juv. is very dark and heavily streaked overall, with speckled throat and buff breast-band.

161.8. Maroon-backed Accentor
Prunella immaculata
15cm   p.355
Dark grey and unstreaked, with chestnut back and belly, and striking yellow eye on blackish face. Juv. is browner above and buff below, coarsely streaked overall with a dark eye, but with adult wing pattern.

PLATE 162. BROWNISH BUNTINGS (L. McQueen)

Buntings *Melophus, Emberiza* (continued on Plate 163)

Breeding males (worn plumage) all have diagnostic head patterns; female, winter male (fresher plumage) and immature usually similar but duller and often hard to identify. Most are found in fairly dry open habitats.

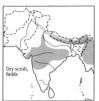

162.1. Crested Bunting
Melophus lathami

15cm p.552
Dark with long crest, rufous in wings. Male black with chestnut wings and tail; fresh male has buffy tips; female brownish-grey and streaked.

162.8. White-capped Bunting
Emberiza stewarti

17cm p.554
Chestnut on scapulars and rump; white outertail feathers. Breeding male is chestnut with white head, and black eyeline and throat; when fresh, mantle is greyer and streaked, with dilute chestnut below. Female pale, with weakly marked greyish head.

162.2. Pallas's Bunting
Emberiza pallasi 15cm p.557

Hypothetical; Nepal. From Reed by smaller bicolored bill, less rufous wings with whiter wing-bars and grey shoulder, and broader pale edges on tail.

162.9. Chestnut-eared Bunting
Emberiza fucata

15cm p.554
Mostly greyish bill. Male has grey crown, chestnut cheek, and rufous on scapulars, rump and side of breast, and streaked necklace; female is duller. Juv. has brown cheek with a heavy dark border and small pale ear-spot, finely streaked breast and flanks.

162.3. 'Thick-billed' Reed Bunting
Emberiza schoeniclus pyrrhuloides

16cm p.557
Paler and larger than Reed with huge bill and little rufous in wing-coverts.

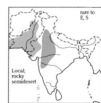

162.10. Striolated Bunting
Emberiza striolata

14cm p.554
Small and pale with bicoloured bill, mottled grey hood with strong head-stripes, pinkish-buff belly, and chestnut shoulder; larger Rock has stronger streaks on mantle. Female and juv. have weaker head pattern.

162.4. Reed Bunting
Emberiza schoeniclus pallidior

15cm p.557
Mid-sized with small bill, blackish patches on sides of throat, pale collar and submoustachial, rufous in wing, and white outertail and belly. Male has black crown, face and throat, obscured in fresh plumage. Female has black malar leading to fine streaks on breast, finely streaked crown and brown cheek.

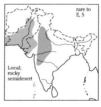

162.11. Rock Bunting
Emberiza cia par

15cm p.553
NW race *par* as next but all plumages much paler.

162.5. Corn Bunting
Miliaria calandra

18cm p.558
Large with swollen bill; drab and streaked, and outertail almost lacks white. See larks (Pl. 97) and female rosefinches (Pl. 167–168).

162.12. *Emberiza cia stracheyi*

Large with long, white-sided tail; pale-headed with strong facial stripes, rufous belly and rump. Breeding male has streaked, rufous upperparts and chestnut underparts. Female paler with features less sharp. Juv. is brown and streaked, with dull rufous rump, plain buffy lower underparts (more uniform than other juvenile buntings), and weak head pattern.

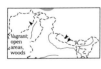

162.6. Rustic Bunting
Emberiza rustica

15cm p.557
Hypothetical; N. Slight crest; broad pale brow, submoustachial and throat, and chestnut streaks on breast and flanks. Male chestnut above with black-and-white head; female and fresh male have head pattern somewhat obscured.

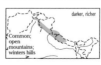

162.7. Little Bunting
Emberiza pusilla

14cm p.555
Small, tiny-billed, and heavily streaked with white outertail feathers; note rufous face and crown, and dark crown- and postocular stripe. Female duller, and pattern obscured when fresh.

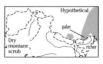

162.13. Godlewski's Bunting
Emberiza godlewskii

15cm p.553
Hypothetical; NE. Like Rock, but has chestnut (not black) crown- and eye-stripes, brighter chestnut scapulars; see text.

PLATE 163. YELLOWISH BUNTINGS (L. McQueen)

Buntings *Emberiza* (continued from Plate 162)

Vagrant; open country

163.1. Yellowhammer
Emberiza citrinella

17cm p.552
Large with longish tail and rufous rump and wing-coverts. Breeding male has yellow head and underparts, and chestnut-streaked upperparts; obscured when fresh, with olive breast-band. Female duller and more streaked, with whiter centre of belly; first-winter female even paler, but still yellowish below.

Common; breeds steppe, winters plains; fields

163.6. Red-headed Bunting
Emberiza bruniceps

17cm p.556
Smaller-billed than often similar Black-headed, with no chestnut on lower mantle and little cheek/throat contrast. Male yellow with streaked olive mantle and chestnut face and breast, much obscured when fresh. Female very like female Black-headed, with buffier cheek, more greenish-yellow rump, and less yellow underparts, often with whiter throat than belly.

Common in E; plains; fields

163.2. Yellow-breasted Bunting
Emberiza aureola

15cm p.555
Pale lower mandible, yellow underparts, and broad white wing-bar. Breeding male chestnut above with black mask, dark breast-band; female with striped head, chestnut rump; juv. paler.

163.3. Pine Bunting
Emberiza leucocephalos

17cm p.552
White belly and crescent on upper breast, with rufous on rump, breast and flanks. Breeding male has chestnut face and throat with white coronal and cheek-stripes; somewhat obscured when fresh. Female and juv. have more diffuse pattern, with darker, heavier streaks below.

Fields, scrub in hills

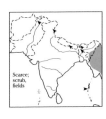

Scarce; scrub, fields

163.7. Chestnut Bunting
Emberiza rutila

14cm p.555
Small with rufous rump and tertials, pale yellow belly, and little white in tail. Male with chestnut hood and upperparts; scaled with buffy edges when fresh. Female olive-brown and streaked above with chestnut rump; first-winter more heavily marked.

163.8. Cinereous Bunting
Emberiza cineracea 17cm p.556

Hypothetical; NE Afghanistan. Plain pale yellowish, with outertail half white. Male with pale greenish-yellow head and underparts, scarcely streaked grey-brown upperparts, and dark wings. Female very finely streaked overall; juv. similar with more streaks above.

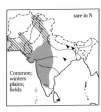

rare in N

Common; winters plains; fields

163.4. Black-headed Bunting
Emberiza melanocephala

18cm p.556
Large with a fairly heavy bill; see Red-headed. Cheek darker than throat, and usually some chestnut on rump. Male with black head and yellow throat and underparts, and unstreaked chestnut back and rump; fresh male has grey crown and black mask. Female very like female Red-headed, but crown more streaked than lower mantle, and cheek contrasts with throat.

rare in S

Semidesert, dry fields

163.9. Grey-necked Bunting
Emberiza buchanani

15cm p.553
Longish pink bill, grey hood, rufous scapulars, grey-brown rump and white outertail. Breeding male with narrow whitish eye-ring and pinkish-rufous underparts; female duller and paler, sometimes finely streaked on crown and mantle. Juv. has streaked crown, spotted breast.

Plains; scrub, grassland

163.5. Black-faced Bunting
Emberiza spodocephala

15cm p.556
Bicoloured bill, yellow belly, and white outertail. Male has black face on olive-grey hood. Female and first-winter have pale yellowish brow, submoustachial and throat, and dappled olive breast-band.

Rare; dry, open scrub

163.10. Ortolan Bunting
Emberiza hortulana

16cm p.554
From Grey-necked by (usually) yellowish eye-ring, throat and submoustachial. Breeding male has unstreaked olive-grey head and breast, greener when fresh. Female has finely streaked olive crown and streaked breast-band. First-winter is brown with heavily streaked upperparts and breast, rufescent rump (unlike juv. Grey-necked) and some chestnut in scapulars, and warm buffy underparts with spotted throat.

PLATE 164. MOUNTAIN-FINCHES AND SNOWFINCHES (L. McQueen)

Mountain-finches *Leucosticte*

Earth-toned, rather large montane finches. Note streaking above, wing pattern.

164.1. Plain Mountain-finch
Leucosticte nemoricola

15cm　　　　　　　　　　p.561
Smaller than Brandt's with heavily streaked mantle, dark rump, pale wing-bars and no white wing-slash; recalls female *Passer* sparrows (Pl. 172). Head is plain brown in NW, streaked on crown with pale grey brow in C and E Himalayas. Juv. has rufescent head and breast and weaker streaks above; first-winter retains unstreaked rufous crown and face through midwinter.

164.2. Sillem's Mountain-finch
Leucosticte sillemi

c.17cm　　　　　　　　　p.561
From Brandt's by cinnamon head without black on forehead; no streaks on mantle, white rump and plainer greyish wings. Juv. much more streaked overall and whiter on belly than juv. Brandt's.

164.3. Brandt's Mountain-finch
Leucosticte brandti

18cm　　　　　　　　　　p.561
Rather large, greyish-buff, and hardly streaked, with blackish forecrown and lores, and white trim on secondaries forming a slash. Breeder has black bill and face, pale pink rump; fresh bird is paler, buffier with pale bill, less black on head. Juv. buffy overall, much paler than juv. Plain.

Snowfinches *Pyrgilauda, Onychostruthus, Montifringilla*

Pale high-mountain finches with fluffy plumage and diagnostic head, wing and tail markings. Juvs. have obscure head patterns, buffy wing- and tail-panels. Formerly all placed in *Montifringilla*.

164.4. Rufous-necked Snowfinch
Pyrgilauda ruficollis

15cm　　　　　　　　　　p.578
Bill longish and mantle streaked; note white face and throat with long black eye-stripe and fine whisker-mark, rufous half-collar and streaked mantle. Juv. plainer, with whisker-mark and white throat.

164.7. White-winged Snowfinch
Montifringilla nivalis

17cm　　　　　　　　　　p.577
Large and rather long-billed, with black eyeline, throat and rump and largely white wing (always more than Black-winged). Breeding male has dark head with white submoustachial; female is paler. Fresh male has obscured rump- and throat-patches, and is less streaked above. Juv. pale and buffy, with weak head pattern, but has diagnostic wing pattern.

164.5. Plain-backed Snowfinch
Pyrgilauda blanfordi

15cm　　　　　　　　　　p.578
Bill stubby and mantle unstreaked; white face with short black eyeline, throat, and coronal stripe, rufous neck-patch, and white wing-slash. Juv. has dilute face pattern.

164.8. Afghan Snowfinch
Pyrgilauda theresae

13cm　　　　　　　　　　p.578
Small with streaked mantle, white upper wing-bar and white patch in flight feathers; less white in tail than other snowfinches. Male has small dark mask, weaker in female. Juv. buffier.

164.6. White-rumped Snowfinch
Onychostruthus taczanowskii

17cm　　　　　　　　　　p.577
Large and very pale with streaked mantle; note white brow over black eyeline, white rump and double wing-bars, white panel and terminal band on secondaries, and broad white corners on longish tail.

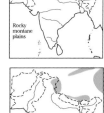
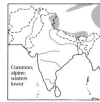

164.9. Black-winged Snowfinch
Montifringilla adamsi

17cm　　　　　　　　　　p.577
Mid-sized, with white carpal patch and greater coverts; less white in wing than similar White-winged, and rump paler. Male has partially obscured dark throat and no dark eyeline; pattern weaker still in female. Juv. has buffier wing-slash.

PLATE 165. CARDUELINE FINCHES (L. McQueen)

Linnet, Twite *Acanthis*

Brownish open-country carduelines with white-edged dark primaries and outertail feathers.

165.1. European Linnet
Acanthis cannabina

13cm   p.560
Chunkier, heavier-billed and shorter-tailed than Twite, with greyish head, streaked throat, and white rump and trim on tail and primaries. Worn male has reddish cap, breast and flanks and chestnut mantle, all obscured in fresh plumage. Female browner and streakier; juv. with buffy wing-bars.

165.2. Twite
Acanthis flavirostris

13cm   p.560
Small with a small stubby bill and long notched tail (usually with little white); drab and streaked except for buffy throat (unlike female Linnet). Breeding male has dark bill and pink rump; paler bill and dull rump in fresh plumage. Female and first-winter have streaked brown rump; similar female Himalayan Beautiful Rosefinch (Pl. 167) is more crisply streaked, especially on face and throat.

Serins, siskins, greenfinches *Carduelis, Serinus, Callacanthis*

Small finches with yellowish patches in wings and tail.

165.3. Himalayan Greenfinch
Carduelis spinoides

14cm   p.559
Male dark above and on cap, cheek and malar, and yellow below, with yellow patches on wing, rump and tail; rather large pale pink bill. Female duller. Juv. entirely streaked, with yellow patches in dark wing and tail. NE race has black face and back and yellow brow.

165.4. Black-headed Greenfinch
Carduelis ambigua

>14cm   p.559
Hypothetical; NE. Like Himalayan, but male has black head, vaguely streaked olive body, broader golden wing-band, and less yellow in tail; female has more olive, streaked head, paler underparts.

165.5. Fire-fronted Serin
Serinus pusillus

12cm   p.558
Heavily streaked above. Worn adult has blackish hood and reddish forehead, both obscured in fresh buffy plumage. First-year has rufous face; juv. is warm brown and plain-breasted.

165.6. Eurasian Siskin
Carduelis spinus

12cm   p.559
Yellow wing-bars and rump. Male has black cap and chin, yellow face and breast, and olive mantle. Female is much plainer, heavily streaked overall; see juv. Himalayan Greenfinch.

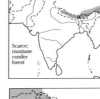

165.7. Tibetan Siskin
Carduelis thibetana

12cm   p.559
Tiny and fine-billed, with yellowish wing-bars and trim but no yellow patches in wings or tail. Male is unstreaked yellow-olive above, yellow below; female duller and more streaked.

165.8. Eurasian Goldfinch
Carduelis carduelis

14cm   p.560
Pale with longish pale bill, red face, broad yellow wing-stripe on black wing, and white rump. First-winter lacks red face; juv. similar but mottled on foreparts.

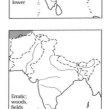

165.9. European Greenfinch
Carduelis chloris

15cm   p.560
Thickset and pale, with stout pale bill, yellow panels in wings and tail. Male is unstreaked yellow-green with yellow face and breast, grey crown and flanks. First-winter female brown with yellow in wing, vaguely streaked above and on breast.

165.10. Red-browed Finch
Callacanthis burtoni

17cm   p.561
Brown with pale bill, white-spotted black wing, white-edged black tail. Male has red 'spectacles' on black head. Female paler, with yellow spectacles, and dusky head. Juv. plainer brown with weak pattern.

PLATE 166. FRINGILLIDS AND DESERT FINCHES (L. McQueen)

Fringillids *Fringilla*

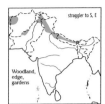

166.1. Common Chaffinch
Fringilla coelebs
15cm   p.558
Buff-brown with bold white wing-bars and white tail-sides. Fresh male has brown crown, buff eyebrow, pinkish underparts. Female and immature drabber but with dark lateral crown-stripe.

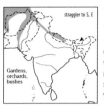

166.2. Brambling
Fringilla montifringilla
15cm   p.558
Orange breast, scapulars and wing-bar, white rump and dark mantle. Breeding male has black head and mantle, obscured in fresh male; female drabber, with dark crown-stripe.

Desert finches *Carpodacus, Bucanetes, Rhodospiza, Rhodopechys*

Desert-adapted species with generally pale coloration but otherwise disparate attributes, grouped here for convenience.

166.3. Sinai Rosefinch
Carpodacus synoicus
15cm   p.566
Very pale and almost plain sandy above, with pale bill. Male has bright crimson face and pink rump and underparts; female plain sandy-buff, vaguely streaked on head and breast.

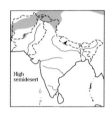

166.6. Mongolian Finch
Bucanetes mongolicus
15cm   p.562
Like Trumpeter but with vague mantle streaks, and white belly. Fresh male washed pale pink, with large white wing-patches; worn male has bright pink patches but yellowish bill. Female duller, and juv. buffier.

166.4. Trumpeter Finch
Bucanetes githagineus
15cm   p.562
Chunky and plain sandy, with stubby bill; lacks white in wing. Fresh male has pale pink-washed plumage; worn male has orange-red bill and brighter pink patches. Female duller. Juv. bright buff overall.

166.7. Crimson-winged Finch
Rhodopechys sanguineus
18cm   p.562
Chunky and mottled, with large yellowish bill. Worn male has black cap and pink face, wings and rump. Female similar, with brown crown and very little pink.

166.5. Desert Finch
Rhodospiza obsoleta
15cm   p.562
Pale and unstreaked, with black spots on pink and white wing, and white-sided black tail. Breeding male has black bill and face; female is plainer, with paler face. Bill pale in non-breeding plumage.

PLATE 167. ROSEFINCHES I (L. McQueen)

Rosefinches *Carpodacus* (also on Plates 166, 169) and *Pyrrhospiza*

Diverse group of montane finches; males largely red or pink, females and immatures brown, more or less streaked, and more difficult to identify. Note especially facial pattern, degree of streaking above, rump colour and wing pattern; also bill size and structure.

167.1. Common Rosefinch
Carpodacus erythrinus roseatus group

15cm p.563

Mid-sized with pale swollen bill, almost unstreaked upperparts, pale double wing-bars and no distinct brow. Male has red head (usually with brownish cheek), breast, flanks and rump, and whitish belly; richer red when breeding (especially in C and E Himalayan *roseatus*), dull rosy to vinous-pink in non-breeding. Female quite plain and greyish with two pale wing-bars, whitish below with weak streaks on breast and flanks; when fresh may be buffier.

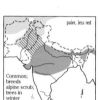

167.2. *Carpodacus e. erythrinus* group

Paler overall, with red less extensive on underparts.

167.3. Himalayan Beautiful Rosefinch
Carpodacus pulcherrimus

15cm p.564

Small, pale and sharply streaked, with longish notched tail and small bill. Male silvery-pink, with heavily streaked flanks. Female heavily streaked overall, and lacks buffy tones (see Twite, Pl. 165).

167.4. Chinese Beautiful Rosefinch
Carpodacus davidianus

p.564

Hypothetical; N Arunachal (not illustrated). Like Himalayan Beautiful but paler, greyer, more finely streaked above, browner, more pale-edged wing, weak flank streaks, browner squared tail. Male is brighter, richer pink, with broader eyebrow.

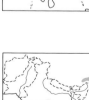

167.5. Pink-rumped Rosefinch
Carpodacus eos

p.564

Possible, N Arunachal (not illustrated). Extremely like Himalayan and Chinese Beautiful but smaller, with finer bill and shorter tail than Chinese; see text.

167.6. Pink-browed Rosefinch
Carpodacus rodochroa

15cm p.565

Small, and streaked above except on rump; see larger, thicker-billed Blyth's. Male pinkish-brown with pink brow, rump and underparts, and dark cap and postorbital patch; female like female Spot-winged (Pl. 169); cheek-patch smaller and not as dark, throat whiter and underparts less heavily streaked.

167.7. Blyth's Rosefinch
Carpodacus grandis

18cm p.566

Big and heavy-billed. Male has vinous-washed mantle, and silvery-pink brow, cheek and throat (*vs* uniformly pink in Pink-browed); female greyish and heavily streaked overall.

167.8. Red-fronted Rosefinch
Pyrrhospiza punicea

20cm p.567

Large with a longish bill and dark, diffusely streaked upperparts. Male has red (not pink) forehead, brow and cheek, and dark brown crown and eyestripe; female almost solid dark above with narrow dark streaks below.

167.9. Streaked Great Rosefinch
Carpodacus rubicilloides

19cm p.567

Large, and darker than Spotted Great. Male strongly streaked above; face and breast crimson with small white spots. Female like female Red-fronted but bill much more conical, with weaker supercilium than female Blyth's.

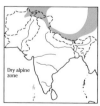

167.10. Spotted Great Rosefinch
Carpodacus severtzovi

19cm p.567

Large and pale, back nearly unstreaked. Male pale pink with large whitish spots. Female very pale and lightly streaked.

PLATE 168. ROSEFINCHES II (L. McQueen)

Rosefinches *Carpodacus* (continued from Plates 166, 167), *Uragus*

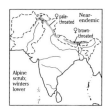

168.1. Himalayan White-browed Rosefinch
Carpodacus thura

17cm p.566
Mid-sized with prominent pale brow. Male has broad pink brow, white at nape, and bright pink rump and underparts; W *blythii* much paler above than E nominate. Female has heavily streaked cheek and underparts; *blythii* with buff brow and breast, nominate with white brow and rusty-brown breast.

168.2. Chinese White-browed Rosefinch
Carpodacus dubius

17cm p.567
Hypothetical; Arunachal (not illustrated). As Himalayan Beautiful but bill smaller, wings and tail longer. Male has whiter eyebrow, more pink on face, and darker underparts. Female has black-streaked, white throat and breast; lacks breast-patch.

168.3. Sharpe's Rosefinch
Carpodacus verreauxii

15cm p.566
Hypothetical; NE. Smaller and slimmer than Spot-winged, with smaller pale bill, weaker eye-stripe and narrower brow nearly reaching bill. Male is darker above than male Spot-winged with more contrasting, paler pink rump, mantle-stripes and underparts, and brownish wing-bars without pink spots. Female is greyer than female Spot-winged, more boldly streaked above, including rump, and slightly paler below.

168.4. Spot-winged Rosefinch
Carpodacus rodopeplus

16cm p.565
Larger than Sharpe's. Male is darker pink on rump and underparts than male Sharpe's, with larger, pinker wing markings. Female like female Sharpe's with bolder buffy brow, darker cheek and almost unstreaked buffy rump; more heavily streaked than smaller female Pink-browed (Pl. 167), with larger, darker cheek-patch, and pale spots on wing-coverts and tertials.

168.5. Dark-rumped Rosefinch
Carpodacus edwardsii

17cm p.565
Stout-billed and dusky with a dark breast and rump. Male muddy vinous-brown with pale pink brow and throat, and pale vinous spots on wing-coverts and tertials. Female duskier brown on breast than female Spot-winged, Himalayan White-browed or Pink-browed, with less conspicuous brow and streaks on breast.

168.6. Dark-breasted Rosefinch
Carpodacus nipalensis

15cm p.563
Dark and stout. Male has pink brow and throat with blackish face mask and breast-band; female brown (dark in E, paler in W), unstreaked below, with buffy wing-bars.

168.7. Vinaceous Rosefinch
Carpodacus vinaceus

13cm p.564
Male dark red with pink brow and white tertial spots; female plain warm brown, lightly streaked below.

168.8. Blanford's Rosefinch
Carpodacus rubescens

15cm p.563
Dark, with longish, fine pale bill. Male variably washed brick-red over grey (see male Common, Pl. 167); female plain with bronzy- or orange-tinged rump, and lacks strong wing-bars.

168.9. Three-banded Rosefinch
Carpodacus trifasciatus

17cm p.565
Hypothetical; Bhutan. Rather large with large pale bill, white scapular line and two wing-bars. Male has pink cowl and flanks, white-streaked face and white belly; female like female Crimson-browed Finch (Pl. 169) but with whiter belly and wing-bars; subadult male has red-tinged brow, breast and rump.

168.10 Long-tailed Rosefinch
Uragus sibiricus

p.568
Possible; N Arunachal (not illustrated). Small, with small pale bill, longish tail. Pale sandy above, with much white in wing and tail. Male has red and pink foreparts and rump, female streaky brown with plain head and cinnamon-tinged rump.

PLATE 169. RED FINCHES, GOLD-NAPED FINCH AND BULLFINCHES (L. McQueen)

Miscellaneous reddish finches Loxia, Propyrrhula, Haematospiza

Bullfinches Pyrrhula

Stubby-billed, plump finches, usually with a small black mask, white band on rump, and extensively black wings and tail with pale greater secondary coverts. Three species are sexually dimorphic (with juveniles resembling the dull females), and have unusually shaped remiges and rectrices. Rather quiet and stolid. Note facial pattern, colour of mantle and nape, pattern of rump.

169.1. Red Crossbill
Loxia curvirostra

15cm   p.568
Chunky with dark wings, shortish notched tail, and heavy bill with crossed tips. Male dark red. Female dark grey tinged green; juv. is greyish with dark streaks overall.

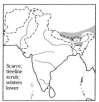

169.2. Crimson-browed Finch
Propyrrhula subhimachala

20cm   p.568
Large, weakly streaked with smallish stubby bill and greyish belly. Male has red foreparts and rump; female has orange-yellow brow and breast, and greenish-gold rump. See Yellow-rumped Honeyguide (Pl. 90).

169.3. Scarlet Finch
Haematospiza sipahi

18cm   p.568
Chunky with stout pale bill. Male glowing red with dark wings and tail. Female is scaly olive-grey with yellow-orange rump; some washed red with orange rump.

Gold-naped Finch Pyrrhoplectes

Rather bullfinch-like but with very different plumage.

169.4. Gold-naped Finch
Pyrrhoplectes epauletta

15cm   p.570
Small, dark and unstreaked, with stubby black bill and white inner tertial webs making white line on back. Male is black with golden-rufous nape and small rufous patch at side of breast. Female is chestnut-brown, paler and pinker below, with dull olive head. See female Dark-breasted and Blanford's Rosefinches (Pl. 168).

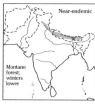

169.5. Red-headed Bullfinch
Pyrrhula erythrocephala

17cm   p.569
Brightest colour on crown and nape, with diffuse white border to black mask, and white rump-patch not bordered above by black. Longer-tailed than Orange, with browner mantle, and shorter-tailed than Grey-headed, with glossier wings and tail. Male has bright orange crown, nape and breast, bronzy in subadult. Female has greenish-yellow nape and grey underparts. In extreme NW, see text for possible Eurasian *P. pyrrhula* (p.569).

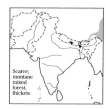

169.6. Grey-headed Bullfinch
Pyrrhula erythaca

17cm   p.569
Grey crown and nape with broad white border to black mask, small white rump-patch below black band, and relatively long tail. Male has grey mantle and orange breast and flanks, more ochraceous in subadult male. Female has pale brown underparts and mantle; less ochraceous below than female Orange.

169.7. Orange Bullfinch
Pyrrhula aurantiaca

14cm   p.569
Small and short-tailed; no white around black mask. Male is bright orange; subadult male is yellower. Female has grey nape and ochre-brown mantle, and is yellower on belly than Grey-headed, with more black on forehead.

169.8. Brown Bullfinch
Pyrrhula nipalensis

17cm   p.569
Large and long-tailed; grey-brown with scaly crown, white patch below eye, and relatively large pale bill.

Plate sponsored by Mr. Perry Bass.

PLATE 170. GROSBEAKS (L. McQueen)

Grosbeaks *Coccothraustes, Eophona, Mycerobas*

Immensely thick-billed finches; males of most are brightly coloured with showy patterns, and females are plainer. Note wing pattern, colour of breast and mantle.

170.1. Hawfinch
Coccothraustes coccothraustes

18cm   p.570

Stout and tawny with small black mask, white shoulder and broad white tip on short tail; huge bill is black in breeding, pale in non-breeding. Female is browner with weaker mask. In flight from above, white coverts and white band through primaries; from below, white wing-lining and primary bases with dark flight feathers.

170.4. Collared Grosbeak
Mycerobas affinis

22cm   p.571

Male very similar to male Black-and-yellow, but richer yellow with orange-tinted nape and mantle, and glossier black head, wings and tail. Yellow thighs diagnostic (if visible). Female is yellow-olive with grey hood and blackish rear mantle, primaries and tail; much more colourful than female Black-and-Yellow. Subadult male duller than adult, so closer to male Black-and-Yellow.

170.2. Chinese Grosbeak
Eophona migratoria

20cm   p.570

Possible in NE. Greyer than Hawfinch, with white-tipped black wings, longish notched black tail and orange bill with black tip; male has black crown, face and throat. In extreme NE, see text for similar hypothetical Japanese *E. personata* (p. 570), less likely for region.

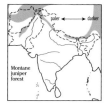

170.5. White-winged Grosbeak
Mycerobas carnipes

22cm   p.571

Large, dark and rather long-tailed with dusky bill and large white carpal spot. Male is black with yellow-olive belly, rump, wing-bar and tertial-spots; black breast diagnostic. Female much duller, grey in place of black.

170.3. Black-and-yellow Grosbeak
Mycerobas icterioides

22cm   p.571

Large with pale bill (darker greenish in summer), straighter at base of culmen than is Collared's. Male is yellow with slaty black hood, wings and tail. Purer yellow below than male Collared, with little or no orange tint above; diagnostic black 'thighs' are very hard to see. See subadult male Collared. Female is plain pale brownish-grey with pale buffy rump and belly; subadult male is similar with blackish wings and yellow rump.

170.6. Spot-winged Grosbeak
Mycerobas melanozanthos

22cm   p.571

Short-tailed with massive, rather rounded dark bill, and white wing-bars, carpal patch and spots on secondaries. Male has black upperparts and throat with lemon-yellow underparts. Female has yellow brow and underparts with broad black eye-stripe, whisker-mark and malar line, and coarse streaks below. Juv. is similar but whitish instead of yellow; subadult male may have almost entirely black head.

Plate sponsored by Mr. Perry Bass.

PLATE 171. AVADAVATS AND MUNIAS (L. McQueen)

Avadavats *Amandava*

Tiny and slim, brightly coloured estrildids with small red bills.

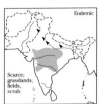

171.1. Green Avadavat
Amandava formosa

10cm p.572

Bright yellow-green above with red bill, and yellow below with barred flanks; female slightly duller. Juv. lacks buffy markings of juv. Red.

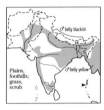

171.2. Red Avadavat
Amandava amandava

10cm p.572

Breeding male red with white spots. Female drab but bill and rump red; eclipse male (illustrated) as female but has brighter rump, larger spots, blacker wings and tail. Juv. plain brown with buffy wing-bars.

Munias *Lonchura, Euodice, Padda*

Small but sturdy, triangular-billed estrildid finches with whirring flight; striking patterns in earth tones. Note especially underpart pattern, presence and position of rump-patch.

171.3. Chestnut Munia
Lonchura atricapilla

10cm p.574

Chestnut with black hood and belly and large silver-blue bill. Juv. buff-brown with slightly grey-hooded appearance.

171.7. Lonchura s. striata

As previous, but peninsular adult *striata* is blackish with white belly. Juv. *striata* very like adult.

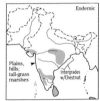

171.4. Tricoloured Munia
Lonchura malacca

10cm p.574

Like Chestnut, but breast and flanks white. Juv. lacks hooded appearance of juv. Chestnut and is all pale below.

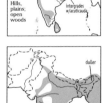

171.8. Scaly-breasted Munia
Lonchura punctulata

10cm p.573

Nut-brown head and upperparts with heavy dark bill and dense black scales on white underparts. Juv. rich buff, paler when worn.

171.5. Indian Silverbill
Euodice malabarica

10cm p.572

Small, slim and pale with large silvery bill, whitish face and throat, white rump and long pointed black tail.

171.9. Black-throated Munia
Lonchura kelaarti

10cm p.573

Black face and throat, pinkish-buff sides of neck and underparts. Peninsular race *jerdoni* has plain underparts and rump; Sri Lanka nominate has black-and-white spotted underparts and rump. Juv. *jerdoni* is plain brown below with whitish shaft-streaks; juv. nominate is dark and scaly, especially on throat.

171.6. White-rumped Munia
Lonchura striata (part)

10cm p.573

Dark brown with white lower back-patch and belly, wedge-shaped tail. Andamans and Nicobars buffier below with scaly breast; N *acuticauda* paler above and heavily marked below. Juv. *acuticauda* mottled overall.

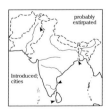

171.10. Java Sparrow
Padda oryzivora

15cm p.574

Pale grey with large rose-red bill, black crown and throat around white cheek. Juv. drab brown with dark bill and pale cheek.

PLATE 172. SPARROWS (L. McQueen)

Sparrows *Passer*

Chunky sparrows with streaked mantles; breeding males have grey and/or chestnut crowns, black lores and bib, streaked mantles, and (usually) black bills; winter males have buffy tips to feathers partly obscuring pattern, and paler bills. Females and juveniles usually plain brownish-grey with pale brow and wing-bars; compare with mountain-finches (Pl. 164). Note especially head pattern and rump colour.

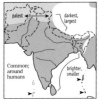

172.1. House Sparrow
Passer domesticus

15cm p.574
Breeding male has grey crown and rump, pale cheek, black bib and dark chestnut nape; fresh male paler and edged with buff. Female plain, with moderate eye-stripe and mantle-streaks; no cheek/throat contrast.

172.2. Eurasian Tree Sparrow
Passer montanus

15cm p.576
Sexes similar; black cheek-spot, small black bib, and solid chestnut crown. Juv. paler, with vague cheek-spot, dark throat (pale NW extreme depicted).

172.3. Sind Sparrow
Passer pyrrhonotus

12cm p.575
Like House but smaller, especially bill. Male has black throat (not breast), and grey on cheek, nape and mid-mantle. Female greyer than female House, with darker crown, more uniform upperparts, and darker shoulder.

172.4. Cinnamon Sparrow
Passer rutilans

15cm p.575
Male is bright rufous above, especially on crown and rump; fresh male paler. Female has bold head pattern and mantle-stripes; female in S Assam hills much darker. In extreme SE, see text for possible Plain-backed *P. flaveolus* (p. 575).

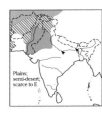

172.5. Spanish Sparrow
Passer hispaniolensis

15cm p.575
Large and heavy-billed. Breeding male has dark chestnut crown and nape, black breast and heavy black streaks on flanks and mantle; fresh and especially juv. male has obscure pattern but diagnostic dark flank markings. Female like smaller female House, but with blurry streaks below and bolder mantle-stripes.

172.6. Dead Sea Sparrow
Passer moabiticus

12cm p.575
Small with well-streaked mantle; male has grey crown, nape and cheek, and chestnut wing-coverts. Female has yellowish bar on neck sides, weak brow and grey cheek.

172.7. Saxaul Sparrow
Passer ammodendri

17cm p.575
Hypothetical; extreme NW. Large and pale with cinnamon side of head and sharp black streaks on mantle. Breeding male has black crown, eye-stripe, throat and shoulder; these are obscure when fresh. Female is plain with slightly streaked crown and whitish wing-bars.

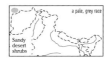

172.8. Desert Sparrow
Passer simplex

p.576
Possible; extreme NW (not illustrated). Very pale and unstreaked, male with black eye-patch and throat, cap solid grey; much white in wings and tail; female plain, paler than similar species.

Petronias *Petronia*

Like *Passer* sparrows, but sexes similar, seasonal plumages differ less, and have a small yellow throat-spot (hard to see in field).

172.9. Yellow-throated Sparrow
Petronia xanthocollis

14cm p.576
Smallish, drab grey; recalls female House Sparrow but unstreaked, with double white wing-bar, and finer black (breeding) or pale (winter) bill. Chestnut shoulder hard to see.

172.10. Rock Sparrow
Petronia petronia

17cm p.576
Chunky with stout pale bill; drab brown and streaked, with dark lateral crown-stripe and eyeline, long broad pale brow, and white tail-spots.

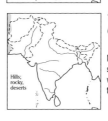

172.11. Pale Sparrow
Carpospiza brachydactyla

p.577
Not illustrated. Very drab, unstreaked sandy overall, with curved culmen, weak head markings, long wings with two white bars, and white tail-tip spots.

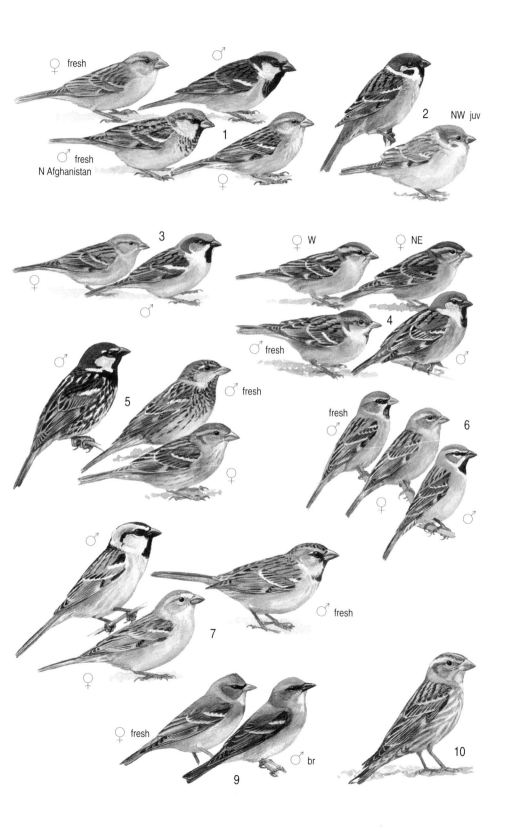

PLATE 173. WEAVERS (L. McQueen)

Weavers or bayas *Ploceus*

Stocky and large-billed; breeding males have yellow crown at least; other plumages brown and streaky. In flocks; build woven nests in colonies. Note especially head, breast pattern.

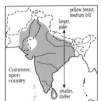

173.1. 'Indian' Baya Weaver
Ploceus p. philippinus

15cm   p.579
Breeding male has yellow cap and breast enclosing black face and throat; female is pale, unstreaked and often washed yellowish below. Juv. bright buff with dusky cheek enclosed by pale brow and submoustachial. In extreme SE, see text for possible Asian Golden *P. hypoxantha* (p. 580).

173.2. 'Eastern' Baya Weaver
Ploceus philippinus burmanicus

Large-billed. Breeding male has golden cap with pale greyish face and throat, and buff underparts; female buff, finely streaked on breast-sides. Juv. bright buff.

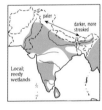

173.3. Streaked Weaver
Ploceus manyar

15cm   p.579
Heavily streaked above and on breast. Breeding male has blackish head with yellow cap; female with yellow-bordered dark cheek, and short dark malar. Juv. pale buff, with pale-scaled crown, greyish cheek and fine triangular breast-streaks (bird illustrated here is least-streaked extreme).

173.4. Finn's Weaver
Ploceus megarhynchus

16cm   p.279
Large and long-tailed with heavy bill and legs. Breeding male has blackish bill, yellow head and breast with dark brown cheek, breast-patch and upperparts; W race with yellow rump and belly, E race with brown rump and white belly. Breeding female paler. Non-breeding adult (both races) dull grey-brown with plain brown head, breast and rump, and broad dark streaks on mantle.

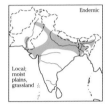

173.5. Black-breasted Weaver
Ploceus benghalensis

15cm   p.278
Small and pale-billed, with (usually) pale throat and some blackish on breast (especially sides). Breeding male has yellow cap, black breast-band, white or (usually) dark cheek. Other plumages have dark breast-patches, blackish cap and eye-stripe, and yellowish eyebrow, submoustachial and throat.

Fody *Foudia*

Like a small weaver but breeding males with red in plumage.

173.6. Madagascar Fody
Foudia madagascariensis   p.580

Chagos; see text (not illustrated). Breeding male bright red; otherwise olive-brown streaked above.

PLATE 174. ORIOLES (J. Anderton)

Orioles *Oriolus*

Stout birds with pointed bills and short legs; males boldly patterned in black and yellow or red. Females similar but duller; immatures streaked below. Note facial pattern, colour of mantle.

Rare; trees, gardens

174.1. **European Golden Oriole**
Oriolus oriolus

25cm   p.586
Vagrant. Very like Indian Golden, but has shorter, darker red bill and much less yellow in wing. Male lacks black behind eye, and has basally black outer rectrices. Female and juv. less sreaked below than same-plumaged Indian, often much greyer on throat.

Hills, plains woods

174.2. **Indian Golden Oriole**
Oriolus kundoo

25cm   p.586
Male yellow with mostly black wings, narrow black mask reaching behind eye, reddish bill, bright red eye and all-yellow outer rectrices; female is duller, with dusky streaks below. Juv. whitish below with blackish streaks and blackish bill.

Hill forest, adjacent plains

174.3. **Maroon Oriole**
Oriolus traillii

28cm   p.587
Very dark, with silvery bill, pale eye, and maroon-red tail. Male purplish-crimson with black head and wings. Female browner with blackish streaks on whitish underparts; throat may be blackish.

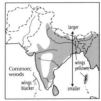

Common; woods

larger
wings yellower
wings blacker
smaller

174.4. **Black-hooded Oriole**
Oriolus xanthornus

25cm   p.587
Yellow with black head and breast and yellow terminal tail-band. Juv. has yellowish forehead and black-streaked throat.

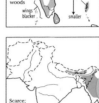

Scarce; hills, lowlands; woods

174.5. **Slender-billed Oriole**
Oriolus tenuirostris

25cm   p.587
Slightly curved bill slimmer than Black-naped's, and black mask narrower at nape; mantle and rump greenish-yellow in both sexes, and female vaguely streaked on nape and breast. Juv. heavily streaked below and has weak nape-band.

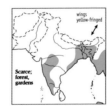

Scarce; forest, gardens

wings yellow-fringed

174.6. **Black-naped Oriole**
Oriolus chinensis diffusus

25cm   p.586
Heavier-billed than Slender-billed, with black mask broader on nape; male has yellow mantle, yellow-olive in female. Juv. very like juv. Slender-billed and Indian Golden, but for heavy bill and (when present) broad nape-band.

Common; forest, gardens

wings blacker
much bigger

174.7. *Oriolus chinensis* (part)

Like previous but richer yellow in Andamans and Nicobars, with wings almost entirely black, and outertail feathers entirely yellow; bill is especially heavy in Nicobars.

Plate sponsored by Mr. Perry Bass.

PLATE 175. DRONGOS (J. Anderton)

Drongos *Dicrurus*

Active, vertical-perching, insectivorous birds, usually glossy black and often with ornamental tail feathers and crests. Many species are very gregarious and vocal, among the most conspicuous members of mixed feeding flocks. Immatures of some species have white-barred belly and vent, and reduced crest and tail features. Note shape of crest and tail, eye colour, pattern of iridescence.

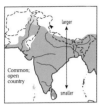

175.1. Black Drongo
Dicrurus macrocercus

31cm   p.590

Semi-glossy black, with long, deeply forked and widely splayed tail; diagnostic small white rictal spot (sometimes absent). First-winter scaled with white on belly and vent.

175.2. Ashy Drongo
Dicrurus leucophaeus

30cm   p.590

Slimmer and less glossy (especially below) than Black, with less splayed tail and brighter red eyes. Widespread nominate dark dull grey below; NE *hopwoodi* monochromatic slate-grey (see Black-winged Cuckooshrike, Pl. 104); NE vagrant *salangensis* much paler with white face-patch.

175.3. Bronzed Drongo
Dicrurus aeneus

24cm   p.591

Small, very glossy and loosely hackled, with heavy forepart-spangling and rather short forked or notched tail. Juv. duller.

175.4. Crow-billed Drongo
Dicrurus annectans

27cm   p.591

Stout and rather plain with thick bill and notched, widely splayed triangular tail less deeply forked than Black. First-winter strongly spangled white below breast.

175.5. White-bellied Drongo
Dicrurus caerulescens

24cm   p.590

Small, dull lead-black with short notched tail, grey throat and breast and white vent; belly extensively white in N, less in south, and grey in S Sri Lanka. Beware first-winter Black.

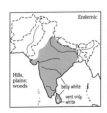

175.6. Lesser Racket-tailed Drongo
Dicrurus remifer

39cm   p.591

Very glossy and flat-headed, with rounded tail and very long streamers with flat rackets.

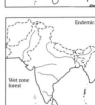

175.7. Ceylon Crested Drongo
Dicrurus lophorinus

31cm   p.593

Large with short nasal crest and deeply forked tail with furled outer feathers but no rackets.

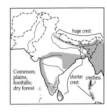

175.8. Greater Racket-tailed Drongo
Dicrurus paradiseus

35cm   p.592

Large, with arching crest and notched tail with long, whorled rackets. Crest shorter in Peninsula, Sri Lanka dry zone and Nicobars, nearly absent in Andamans (figure on plate with one aberrant furled outertail feather; based on specimen). First-winter with scattered white spots below and short crest.

175.9. Hair-crested Drongo
Dicrurus hottentottus

31cm   p.591

Velvety black with glossy wings and breast- and neck-hackles, and triangular tail with furled corners; note decurved, heavy bill. Juv. duller.

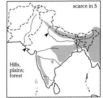

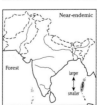

175.10. Andaman Drongo
Dicrurus andamanensis

35cm   p.592

Mid-sized with rather swollen bill and forked tail with furled outer feathers.

175.11. 'Fork-tailed' Drongo-cuckoo
Surniculus [lugubris] dicruroides   27cm   Pl. 72, p.228

Drongo-like but with a fine, decurved bill and white bars on vent and undertail (juv. drongos may have barred vent but not tail); lazy cuckoo posture.

PLATE 176. STARLINGS I (J. Anderton)

Starlings and Mynas Sturnidae

Stocky, noisy birds with short tails and pointed wings. Many are strongly associated with humans; some are gifted mimics.

176.1. Spot-winged Starling
Saroglossa spiloptera

19cm   p.580

Small and brown with yellow eye and small white carpal patch. Male is scaled above, with dark mask, dark chestnut throat, rufous-buff breast and flanks. Female is whitish below with slightly mottled throat and breast.

Starlings *Sturnus, Temenuchus, Sturnia* (and *Gracupica* on Pl. 177)

Chunky, short-tailed gregarious birds with long, pointed bills and usually lanceolate plumage in adults. Vocalisations usually raucous and squealy. Adults easily separated, but plainer juveniles lack hackles and are often similar.

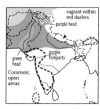

176.2. Common Starling
Sturnus vulgaris

20cm   p.583

Breeding plumage glossy black with pointed yellow bill; in winter heavily spangled with white, bill dusky. Juv. muddy brown, sometimes slightly streaked below, with dark bill.

176.3. Daurian Starling
Sturnia sturnina

19cm   p.582

Vagrant. Small and pale, with a short dark bill, glossy purple nape-patch and mantle, white wing-bars, cinnamon rump, vent and outertail. Juv. duller overall.

176.4. Brahminy Starling
Temenuchus pagodarum

22cm   p.582

Pale brown above; orange-buff below with black cap and crest (usually sleeked down), blue-based yellow bill, white vent and outertail. Juv. similar but crown brown and uncrested, and base of bill pale.

176.5. Rosy Starling
Sturnus roseus

23cm   p.582

Breeding pale pink with black hood, wings, tail and vent, and reddish bill and legs. Fresh adult browner, with scaled hood. Juv. drab, paler brown than juv. Common, with thicker yellowish bill.

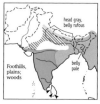

176.6. Grey-headed Starling
Sturnia malabarica

21cm   p.581

Pale grey with whitish eye and hackled head, blue-based yellowish bill and chestnut outertail. Widespread nominate washed rufous below breast; NE *nemoricola* sometimes nearly white below.

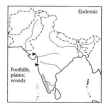

176.7. Malabar White-headed Starling
Sturnia blythii

>21cm   p.581

Like Grey-headed, but male with brighter white hood, more rufous belly; more contrasting soft part colors. See text for female.

176.8. White-shouldered Starling
Sturnia sinensis   20cm   p.582

Hypothetical: Manipur. Small and very pale, with whitish eye, white scapulars and coverts on black wings and broad white tail-edges. Juv. browner with all-dark wings.

176.9. Andaman White-headed Starling
Sturnia erythropygia

21cm   p.581

White with black wings and tail, yellowish bill and legs; little rufous on Andamans; on Car Nicobars, rump and vent chestnut; on C Nicobar, rump pale rufous.

176.10. Vinous-breasted Starling
Sturnia burmannica

22cm   p.582

Possible; extreme E. Pinkish-grey with whitish head; suggests Grey-headed, but larger, with dark eyeline and reddish-based bill.

176.11. White-faced Starling
Sturnia albofrontata

22cm   p.581

Greenish-black above with bold white face. Juv. dull brown overall with pale face.

PLATE 177. STARLINGS II AND MYNAS (J. Anderton)

Miscellaneous sturnids *Gracupica, Aplonis, Ampeliceps*

177.1. Black-collared Starling
Gracupica nigricollis

27cm   p.583
Possible; extreme E. Very large with white head and underparts, and black bill, collar and breast.

177.3. Asian Glossy Starling
Aplonis panayensis

22cm   p.580
Adult entirely glossy green-black with short stout dark bill, and large eyes. Juv. has whitish underparts with bold black streaks.

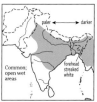

177.2. Asian Pied Starling
Gracupica contra

23cm   p.583
Pied, with red-based yellow bill, red eye-ring in white cheek-patch, and white scapular line, rump and belly. Juv. has black replaced with brown.

177.4. Gold-crested Myna
Ampeliceps coronatus

21cm   p.585
Black with yellow crown, throat and carpal patch; female duller with less yellow. Juv. has black crown.

Mynas *Acridotheres*

Mostly larger and heavier than starlings, usually with stout yellow or orange bill and legs, and bristly frontal or nasal crests, longest in adult males. Generally dark with pale carpal patch and outertail feathers conspicuous in flight. Note colour of bill and orbital skin, belly and vent.

177.5. Bank Myna
Acridotheres ginginianus

21cm   p.584
Grey with black cap, and red-orange bill, orbital skin and legs; note buff belly, wing-patch and outertail. Juv. has brownish head, dull soft parts.

177.8. White-vented Myna
Acridotheres grandis

23cm   p.584
Large with yellow bill, dark eye, frilly frontal crest and clear-cut white vent. Juv. browner with short crest, white-blotched belly.

177.6. Common Myna
Acridotheres tristis

23cm   p.584
Brown with black hood, yellow bill, orbital skin and legs, and white belly and vent. Sri Lanka form *melanosterna* darker, often with more extensive facial skin.

177.9. Collared Myna
Acridotheres albocinctus

23cm   p.584
Whitish half-collar, short frontal crest, and white-scaled vent.

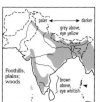

177.7. Jungle Myna
Acridotheres fuscus

23cm   p.584
Greyer than Common with pale eye (no bare skin), short-looking dark-based bill and bristly nasal crest. Juv. browner with pale belly and throat, reduced nasal crest.

PLATE 178. HILL-MYNAS, NUTCRACKERS, JACKDAWS AND CHOUGHS (H. Burn)

Hill-mynas *Gracula*

Large and very stocky glossy black sturnids with orange fleshy wattles on nape, and sturdy yellow legs. Vocally remarkable. Juvs. similar but duller, with less-developed, duller wattles. Note bill colour, wattle shape and calls.

178.1. Common Hill-myna
Gracula religiosa

29cm p.585
Heavy orange bill; cheek- and nape-wattles connected below eye.

178.2. Lesser Hill-myna
Gracula indica

25cm p.585
Smaller and smaller-billed than Common, with long hindcrown-wattle and disjunct cheek-wattle.

178.3. Ceylon Hill-myna
Gracula ptilogenys

25cm p.585
Like sympatric Lesser, but lacks cheek-wattle; eye is usually pale, and bill has blue base.

Nutcrackers *Nucifraga*

Chunky, short-tailed, dark brown montane corvids with distinctive flight patterns.

178.4. Larger-spotted Nutcracker
Nucifraga multipunctata

32cm p.596
Dark with long slim bill, heavy white spots on foreparts, and white rump, vent and outertail.

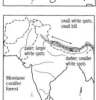

178.5. Spotted Nutcracker
Nucifraga caryocatactes

32cm p.597
Darker, thicker-billed and much less heavily and generally spotted than NW Larger-spotted.

178.6. Pleske's Ground-jay
Podoces pleskei

p.597
Hypothetical; extreme W; see text (not illustrated). Sandy desert bird, with black breast-band, black-and-white wings, black tail.

178.7. Pander's Ground-jay
Podoces panderi

p.597
Possible; extreme NW; see text (not illustrated). Pale grey desert bird, with black breast-band, mainly white wings, black tail.

178.8. Ground-tit
Pseudopodoces humilis 20cm Pl. 155, p.535

178.9. Eurasian Jackdaw
Coloeus monedula

33cm p.598
Small with short stout bill, white eye, and silvery hindcrown, cheek and nape. Juv. duller.

178.10. Daurian Jackdaw
Coloeus dauuricus

p.598
Hypothetical; NE (not illustrated). Like Eurasian but with white collar and belly.

Choughs *Pyrrhocorax*

Black, highly aerial corvids of cliffs and mountains, both with red legs, unlike other corvids. Note bill length, curvature and colour; shape of wings and tail, especially in flight.

178.11. Red-billed Chough
Pyrrhocorax pyrrhocorax

45cm p.597
Glossy black with long, curved red bill, yellow in immature. Flatter-crowned and heavier-legged than Alpine, with longer, deeper-based bill; in flight, more 'fingered' primaries and shorter, squarer tail.

178.12. Alpine Chough
Pyrrhocorax graculus

38cm p.598
From Red-billed by shorter, straighter, yellow bill; wing-tips fall further short of tail-tip.

PLATE 179. CROWS AND RAVENS (H. Burn)

Crows and **ravens** *Corvus*

Medium to very large black or black-and-grey corvids; longer-winged and shorter-tailed than jays. Ravens are largest, with very heavy bills and wedge-shaped tails, although these features are approached in many crows. Note tail shape, bill size and shape, plumage pattern and gloss.

179.1. Rook
Corvus frugilegus

48cm   p.591

Mid-sized and small-headed, with spike-like bill and bare whitish face; note shaggy thighs.

179.6. Eastern Jungle Crow
Corvus [*macrorhynchos*] *levaillantii*

47cm   p.600

Not illustrated. From Indian Jungle by voice; culmen usually less arched but deeper near tip and plumage less purple-glossed.

179.2. Hooded Crow
Corvus [*corone*] *cornix*

<47cm   p.599

Same shape as larger Carrion, but pale grey with black hood, wings and tail. Pale NW race of House has whitish hindcrown and rear cheek, and lacks black breast-patch.

179.7. Large-billed Crow
Corvus [*macrorhynchos*] *japonensis*

50cm   p.599

Large, glossy black with high-arched heavy bill and steep forehead; lacks 'beard' of Common Raven. See Indian and Eastern Jungle.

179.3. Carrion Crow
Corvus corone (part)

>47cm   p.599

Race *orientalis* is mid-sized, with fairly short, deep bill, and flattish forecrown; wings almost reach tip of square tail.

179.8. Brown-necked Raven
Corvus ruficollis

58cm   p.600

Smaller than Common, with less 'bearded' look and browner foreparts, shorter wings and more rounded tail, often with projecting central rectrices.

179.4. House Crow
Corvus splendens

43cm   p.598

Rather lightweight with a longish neck; slimmer-billed and scrawnier than the jungle crows, with less full belly feathering. Smoky collar is obvious in most of range, nearly white in NW (see Hooded), but dark and nearly obsolete in Sri Lanka, where see Indian Jungle.

179.9. Common Raven
Corvus corax tibetanus

69cm   p.600

Very large and 'bearded', with heavy bill with long, thick nasal bristles, and wedge-tipped tail; wings reach tail-tip.

179.5. Indian Jungle Crow
Corvus [*macrorhynchos*] *culminatus*

49cm   p.600

Smallish and glossy with squarish tail; similar to Large-billed, but bill smaller and less arched, forehead less steep, and feet less heavy. From Eastern Jungle by voice. Beware very dark House in Sri Lanka.

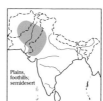

179.10. Punjab Raven
Corvus corax subcorax

<69cm   p.600

Intermediate between Common and Brown-necked, closer to Common; no range overlap.

PLATE 180. MAGPIES, JAYS AND TREEPIES (J. Anderton)

Magpie *Pica*

180.1. Eurasian Magpie
Pica pica

52cm p.596
Black with white belly and scapular patches, long glossy tail, and white 'slots' in primaries prominent in flight. W race longer-tailed and has small white rump-patch, lacking in stockier E form.

Jays *Garrulus*

Corvids with medium-length tails; note wing, rump, head, tail patterns.

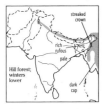

180.2. Eurasian Jay
Garrulus glandarius

33cm p.593
Pinkish-rufous with black bill, moustache, wings and tail, and white rump and vent. Overall colour markedly regionally variable.

180.3. Black-headed Jay
Garrulus lanceolatus

33cm p.594
Greyer than Eurasian with crested black head, blue tail, and no white rump.

Magpies *Cissa, Urocissa*

Spectacular blue or green corvids with long tails and brightly coloured bills and legs; in blue magpies *Urocissa* note head pattern, bill colour.

180.4. Common Green Magpie
Cissa chinensis

47cm p.595
Lime-green with chestnut wings, black mask and red bill and eye-ring.

180.5. Yellow-billed Blue Magpie
Urocissa flavirostris

66cm p.594
Very like Red-billed but yellow bill and more orangey legs, and smaller white nape-patch without spangled crown.

180.6. Red-billed Blue Magpie
Urocissa erythrorhyncha

70cm p.594
Blue with long white-tipped tail, red bill and legs, and black hood with white nape and white-spangled hindcrown.

180.7. Ceylon Blue Magpie
Urocissa ornata

47cm p.594
Deep blue with long white-tipped tail, chestnut hood and wings, and red bill, eye-ring and legs.

Treepies *Dendrocitta*

Earth-toned corvids with long tails and very thick short bills. Note wing and tail pattern.

180.8. Grey Treepie
Dendrocitta formosae

42cm p.595
Dusky grey with weakly defined black face and throat; darker than Collared with white carpal patch on wholly black wings, and two-toned tail.

180.9. Collared Treepie
Dendrocitta frontalis

38cm p.595
Smallish; like Grey but with very pale hindneck and breast contrasting sharply with black face and chestnut back and underparts; note also solid black tail. In extreme SE, see text for possible Hooded *Crypsirina cucullata* (p.596).

180.10. Andaman Treepie
Dendrocitta bayleyi

36cm p.596
Small; dark chestnut with black hood, wings and tail and striking pale eye.

180.11. Rufous Treepie
Dendrocitta vagabunda

50cm p.595
Cinnamon-buff with blackish hood, largely whitish wing and black-tipped grey tail.

180.12. White-bellied Treepie
Dendrocitta leucogastra

50cm p.596
Largely white with black face and wings, rufous mantle and black-tipped tail.

Plate sponsored by Mrs. Evelyn Bartlett.

Plate index to genera and group names

A

Abroscopus 149
Acanthis 165
Accentors 161
Accipiter 22, 23
Aceros 88, 89
Acridotheres 177
Acrocephalus 146, 147
Actinodura 136
Actitis 55
Adjutants 14, 13
Aegithalos 155
Aegithina 108
Aegolius 79
Aegypius 32
Aerodramus 82
Aethopyga 160
Aix 19, 20
Alaemon 97
Alauda 96
Alcedo 86
Alcippe 135
Alectoris 38
Alophoixus 106
Amandava 171
Amaurornis 48, 49
Ammomanes 98
Ammoperdix 38
Ampeliceps 177
Anas 17, 20
Anastomus 12, 13
Anhinga 5
Anous 64
Anser 15
Anthocincla 95
Anthracoceros 88, 89
Anthreptes 159
Anthropoides 45, 46
Anthus 102, 103
Aplonis 177
Apus 83
Aquila 28, 29
Arborophila 39
Ardea 8, 9
Ardeola 10, 9
Ardeotis 47, 46
Arenaria 56
Artamus 104
Asarcornis 16

Asio 77
Athene 79
Avadavats 171
Aviceda 21
Avocet 50
Aythya 18, 20

B

Babaxes *Babax* 128
Babblers 131–139
Bamboo-partridge 39
Bambusicola 39
Barbets 90
Barn-owls 75
Barwings 136
Batrachostomus 80
Bayas 173
Bazas 21
Bee-eaters 87
Besra 22, 23
Bitterns 11, 9
Blackbirds 113
Bluetails 116
Bluethroat 117
Blythipicus 93
Bombycilla 110
Boobooks 75
Booby (Boobies) 6
Botaurus 11, 9
Brachypteryx 115
Bradypterus 145
Brambling 166
Branta 15
Broadbills 95
Bubo 76
Bubulcus 7, 9
Bucanetes 166
Bucephala 19, 20
Buceros 88, 89
Bulbuls 106, 107
Bullfinches 169
Bulweria 3
Buntings 162, 163
Burhinus 51
Bushchats 122
Bush-larks 96
Bush-quails 40
Bush-robins 116
Bush-warblers 144, 145

Bustard-quail 40
Bustards 47, 46
Butastur 25
Buteo 24
Butorides 11, 9
Buttonquail 40
Buzzards 24, 25

C

Cacomantis 72
Cairina 16
Calandrella 98
Calidris 56, 57
Callacanthis 165
Caloenas 67
Calonectris 2
Canary-flycatcher 149
Caprimulgus 80, 81
Carduelis 165
Carpodacus 166–168
Carpospiza 172
Catharacta 59
Catreus 44
Cecropis 100
Centropus 74
Cephalopyrus 156
Cercomela 118
Cercotrichas 117
Certhia 156
Ceryle 86
Cettia 144
Ceyx 86
Chaetornis 141
Chaffinch 166
Chaimarrornis 120
Chalcoparia 159
Chalcophaps 69
Charadrius 53
Chats 115–122
Chelidorhynx 110
Chiffchaffs 150
Chlamydotis 47, 46
Chlidonias 64
Chloropsis *Chloropsis* 108
Choughs 178
Chrysococcyx 72
Chrysocolaptes 94
Chrysomma 131
Chukar 38

Ciconia 12, 13
Cinclidium 115
Cinclus 110
Cinnyris 159
Circaetus 26
Circus 33, 34
Cissa 180
Cisticolas *Cisticola* 141
Clamator 72
Clangula 19, 20
Coccothraustes 170
Cochoas *Cochoa* 111
Collared-doves 68
Collocalia 82
Coloeus 178
Columba 66, 67
Coot 49
Coppersmith 90
Copsychus 118
Coracias 84
Coracina 104
Cormorants 5
Corvus 178, 179
Coturnix 40
Coucals 74
Coursers 51
Crab-plover 50
Crag-martins 99
Crakes 48, 49
Cranes 45, 46
Creepers 156
Crex 48
Criniger 106
Crossbill 169
Crossoptilon 43
Crows 179
Cuckoo-doves 68
Cuckoos 72, 73
Cuckooshrikes 104
Cuculus 73
Culicicapa 149
Curlews 54
Cursorius 51
Cutia *Cutia* 137
Cygnus 15
Cyornis 125, 126
Cypsiurus 82

D

Dabchick 1
Daption 2
Darter 5
Delichon 99
Dendrocitta 180

Dendrocopos 91, 92
Dendrocygna 16
Dendronanthus 101
Dicaeum 158
Dicrurus 175
Dinopium 94
Dippers 110
Divers 1
Dollarbird 84
Doves 68, 69
Dowitchers 54
Dromas 50
Drongos 175
Dryocopus 94
Dryonastes 128–130
Ducks 16–20
Ducula 66
Dumetia 131
Dunlin 56
Dupetor 11, 9

E

Eagle-owls 76
Eagles 25–30
Eared-nightjar 80, 81
Eared-pheasant 43
Egrets *Egretta* 7, 9
Elanus 21
Elaphrornis 145
Emberiza 162, 163
Enicurus 118
Eophona 170
Ephippiorhynchus 12, 13
Eremophila 97
Eremopterix 97
Erithacus 116, 117
Erpornis *Erpornis* 134
Esacus 51
Estrilda 171
Eudynamys 72
Eumyias 126
Euodice 171
Eurostopodus 80, 81
Eurynorhynchus 57
Eurystomus 84
Excalfactoria 40

F

Fairy-bluebird 108
Falco 35–37
Falconets 35
Falcons 36, 37
Fantails 110

Ficedula 123–125
Fieldfare 112
Finchbill 107
Finches 164–169
Finch-larks 97
Finfoot 49
Firethroat 116
Fish-eagles 30
Fish-owls 76
Flamebacks 94
Flamingos 14, 13
Floricans 47, 46
Flowerpeckers 158
Flycatchers 123–126, 149
Flycatcher-shrike 104
Fody 173
Forktails 118
Foudia 173
Francolins *Francolinus* 38
Fregata 6
Fregetta 3
Frigatebirds 6
Fringilla 166
Frogmouths 80
Fulica 49
Fulvettas 135

G

Gadwall 17, 20
Galerida 96
Gallicrex 49
Gallinago 58
Gallinula 49
Galloperdix 41
Gallus 41
Gampsorhynchus 136
Garganey 17, 20
Garrulax 127–130
Garrulus 180
Gavia 1
Gecinulus 93
Gelochelidon 63
Glareola 51
Glaucidium 79
Godwits 54
Goldcrest 148
Goldenbacks 94
Goldeneye 19, 20
Goldfinch 165
Goosander 19, 20
Goose (Geese) 15
Gorsachius 10, 9
Goshawks 22, 23
Gracula 178

Gracupica 177
Graminicola 141
Grammatoptila 130
Grandala *Grandala* 111
Grassbirds 141
Grass-owl 75
Grass-warbler 141
Grebes 1
Greenfinches 165
Green-pigeons 69
Greenshanks 55
Griffons 31
Grosbeaks 170
Ground-jays 178
Ground-thrushes 111, 114
Ground-tit 155, 178
Grus 45, 46
Gulls 60, 61
Gygis 63
Gypaetus 32
Gyps 31

H

Haematopus 50
Haematospiza 169
Halcyon 85
Haliaeetus 30
Haliastur 21
Hanging-parrots 70
Harpactes 84
Harriers 33, 34
Hawfinch 170
Hawk-cuckoos 73
Hawk-eagles 26, 27
Hawk-owls 75
Heliopais 49
Hemicircus 91
Hemiprocne 82, 100
Hemipus 104
Hemixos 106
Herons 7–11
Heteroglaux 79
Heterophasia 136
Heteroscelus 55
Heteroxenicus 115
Hieraaetus 26
Hierococcyx 73
Hill-mynas 178
Hill-partridges 39
Himantopus 50
Hippolais 147
Hirundapus 83
Hirundo 99, 100
Hobby (Hobbies) 36

Hodgsonius 120
Honey-buzzards 25
Honeyguide 90
Hoopoe 84
Hornbills 88, 89
Houbara 47, 46
Houbaropsis 47, 46
House-martins 99
Hydrophasianus 50
Hydroprogne 63
Hypocolius *Hypocolius* 110
Hypopicus 92
Hypothymis 110
Hypsipetes 106, 107

I

Ianthocichla 127, 128
Ibidorhyncha 50
Ibisbill 50
Ibises 14, 13
Ichthyophaga 30
Ictinaetus 27
Imperial-pigeons 66
Indicator 90
Iole 106, 107
Ioras 108
Irania 120
Irena 108
Ithaginis 42
Ixobrychus 11, 9

J

Jacanas 50
Jackdaws 178
Jaegers 59
Jays 178, 180
Junglefowl 41
Jynx 91

K

Kestrels 35
Ketupa 76
Kingfishers 85, 86
Kites 21
Kittiwake 61
Knots 56
Koel 72

L

Laggar 37
Lalage 104

Lammergeier 32
Lanius 109
Lapwings 52
Larks 96–98
Larus 60, 61
Laughingthrushes 127–130
Leafbirds 108
Leaf-warblers 150–152
Leiothrix *Leiothrix* 134
Leptocoma 159
Leptopoecile 148
Leptoptilos 14, 13
Lerwa 39
Leucosticte 164
Limicola 56
Limnodromus 54
Limosa 54
Linnet 165
Liocichla *Liocichla* 128
Locustella 141
Lonchura 171
Loons 1
Lophophorus 42
Lophura 43
Loriculus 70
Lorikeets 70
Loxia 169
Lullula 96
Luscinia 116, 117
Lymnocryptes 58

M

Macronous 131
Macropygia 68
Magpie-robin 118
Magpies 180
Malacocincla 139
Malkohas 74
Mallard 17, 20
Marmaronetta 19, 20
Martins 99
Megalaima 90
Megalurus 141
Megapode *Megapodius* 39
Melanitta 19
Melanochlora 155
Melanocorypha 97
Melophus 162
Mergansers 19, 20
Mergellus 19, 20
Mergus 19, 20
Merlin 35
Merops 87
Mesia 134

Metopidius 50
Microhierax 35
Micropternus 93
Miliaria 162
Milvus 21
Minivets 105
Minlas *Minla* 137
Mirafra 96
Monal-partridge 43
Monals 42
Monarch 110
Monticola 111
Montifringilla 164
Moorhen 49
Motacilla 101
Mountain-finch 164
Moupinia 131
Mulleripicus 94
Munias 171
Muscicapa 123–126
Muscicapella 125
Mycerobas 170
Mycteria 12, 13
Myiomela 115
Mynas 176–178
Myophonus 111
Myzornis *Myzornis* 134

N

Napothera 139
Nectarinia 159
Needletails 82, 83
Neophron 32
Netta rufina 18, 20
Nettapus 19, 20
Night-herons 10, 9
Nightingale 117
Nightjars 80, 81
Niltavas *Niltava* 125, 126
Ninox 75
Noddy (Noddies) 64
Nucifraga 178
Numenius 54
Nutcrackers 178
Nuthatches 157
Nyctea 76
Nycticorax 10, 9
Nyctyornis 87

O

Oceanites 3
Oceanodroma 3
Ocyceros 88, 89

Oenanthe 121
Onychostruthus 164
Openbill 12, 13
Ophrysia 39
Orioles *Oriolus* 174
Orthotomus 148
Ortolan 163
Osprey 30
Otis 47, 46
Otus 78
Ouzel 113
Owlets 79
Owls 75–79
Oxyura 18, 20
Oystercatcher 50

P

Pachycephala 123
Padda 171
Painted-snipe 50, 58
Palm-swift 82
Pandion 30
Panurus 140
Papasula 6
Paradise Flycatcher 110
Paradoxornis 140
Parakeets 70, 71
Parrotbill 140
Partridge 38, 39, 43
Parus 154, 155
Passer 172
Pavo 44
Peacock-pheasant 43
Peafowl 44
Pelagodroma 3
Pelargopsis 85
Pelecanus 4
Pelicans 4
Pellorneum 139
Penduline-tits 156
Perdicula 40
Perdix 38
Peregrine 37
Pericrocotus 105
Pernis 25
Petrels 2, 3
Petronias *Petronia* 172
Phaenicophaeus 74
Phaethon 4
Phalacrocorax 5
Phalaropes *Phalaropus* 57
Phasianus 44
Pheasants 42–44
Philomachus 56

Phodilus 75
Phoeniconaias 14, 13
Phoenicopterus 14, 13
Phoenicurus 119, 120
Phragamaticola 146
Phylloscopus 149–152
Pica 180
Picoides 91, 92
Piculets 91
Picumnus 91
Picus 93
Pigeons 66, 67, 69
Pintail 17, 20
Pipits 102, 103
Pittas *Pitta* 95
Platalea 14, 13
Plegadis 14, 13
Ploceus 173
Plovers 52, 53
Pluvialis 53
Pnoepyga 138
Pochards 18, 20
Podiceps 1
Podoces 178
Polyplectron 43
Pomatorhinus 133
Pond-herons 9, 10
Porphyrio 49
Porzana 48, 49
Pratincoles 51
Prinias *Prinia* 142, 143
Propyrrhula 169
Prunella 161
Psarisomus 95
Pseudibis 14, 13
Pseudobulweria 3
Pseudopodoces 155, 178
Psittacula 70, 71
Pterocles 65
Pterodroma 3
Pteruthius 137
Ptilolaemus 88, 89
Ptyonoprogne 99
Pucrasia 43
Puffinus 2
Pycnonotus 106, 107
Pyrgilauda 164
Pyrrhocorax 178
Pyrrhoplectes 169
Pyrrhospiza 167
Pyrrhula 169

Q

Quails 39, 40

R

Rails 48
Rallina 48
Rallus 48
Ravens 179
Recurvirostra 50
Redshanks 55
Redstarts 119, 120
Redwing 112
Reedling 140
Reed-warblers 146, 147
Reef-herons 7, 9
Regulus 148
Remiz 156
Rhinomyias 123
Rhinoptilus 51
Rhipidura 110
Rhodonessa 18, 20
Rhodopechys 166
Rhodospiza 166
Rhopocichla 131
Rhyacornis 120
Rhyticeros 88, 89
Rimator 139
Riparia 99
Rissa 61
Robins 115–118
Rock-chat 118
Rock-thrushes 111
Rollers 84
Rook 179
Rosefinches 166–168
Rostratula 50, 58
Rubycheek 159
Rubythroats 117
Ruff 56
Rynchops 64

S

Saker 37
Salpornis 156
Sanderling 57
Sandgrouse 65
Sand-martins 99
Sandpipers 55–57
Sarcogyps 32
Sarkidiornis 16
Saroglossa 176
Sasia 91
Saxicola 122
Saxicoloides 118
Scaup 18, 20
Schoenicola 141

Scimitar-babbler 133
Scolopax 58
Scops-owls 78
Scoter 19
Scotocerca 143
Scrubfowl 39
Scrub-robin 117
Sea-eagle 30
Seicercus 149
Serilophus 95
Serin *Serinus* 165
Serpent-eagles 25
Shag 5
Shaheen 37
Shamas 118
Shearwaters 2
Shelducks 16
Shikra 22, 23
Shortwings 115
Shoveler 17, 20
Shrike-babblers 137
Shrikes 109
Sibias 136
Silverbill 171
Sirkeer 74
Siskins 165
Sitta 157
Skimmer 64
Skuas 59
Skylark 96
Smew 19, 20
Snake-eagle 26
Snipes 58
Snowcocks 43
Snowfinches 164
Sparrowhawks 22, 23
Sparrow-larks 97
Sparrows 172
Spelaeornis 138
Sphenocichla 139
Spiderhunters 160
Spilornis 25
Spinetails 82, 83
Spizaetus 27
Spizixos 107
Spoonbill 14, 13
Spurfowl 41
Stachyris 131, 139
Stactocichla 128
Staphida 134
Starlings 176, 177
Stercorarius 59
Sterna 62–64
Stilts 50
Stints 57

Stonechats 122
Stone-curlews 51
Storks 12, 13
Storm-petrels 3
Streptopelia 68
Strix 77
Stubtail 144
Sturnia 176
Sturnus 176
Sula 6
Sunbirds 159
Surniculus 72
Suthoras 140
Swallows 100
Swamphen 49
Swans 15
Swiftlets 82
Swifts 82, 83
Sylvia 153
Sylviparus 156
Sypheotides 47, 46
Syrmaticus 44
Syrrhaptes 65

T

Taccocua 74
Tachybaptus 1
Tachymarptis 83
Tadorna 16
Tailorbirds 148
Tarsiger 116
Tattler 55
Teal 17, 19, 20
Temenuchus 176
Tephrodornis 104
Terns 62–64
Terpsiphone 110
Tesias *Tesia* 148
Tetraogallus 43
Tetraophasis 43
Tetrax 46, 47
Thalasseus 63
Thick-knees 51
Threskiornis 14, 13
Thrushes 111–114
Tichodroma 156
Tickellia 149
Timalia 131
Tit-babblers 131, 135
Tits 154–156
Tit-warblers 148
Todiramphus 85
Tragopans *Tragopan* 42
Treecreepers 156

Treepies 180
Treeswift 82, 100
Treron 69
Trichastoma 139
Triller 104
Tringa 55
Trochalopteron 127–129
Troglodytes 138
Trogons 84
Tropicbirds 4
Tryngites 56
Turdoides 132
Turdus 112, 113
Turnix 40
Turnstone 56
Turtle-doves 68
Twite 165
Tyto 75

U

Upupa 84
Uragus 168
Urocissa 180
Urosphena 144

V

Vanellus 52
Vultures 31, 32

W

Wagtails 101
Wallcreeper 156
Warblers 141, 143–147, 149–153
Watercock 49
Waterhen 49
Waxwing 110
Weavers 173
Wheatears 121
Whimbrel 54
Whinchat 122
Whistler 123
Whistling-ducks 16
Whistling-thrushes 111
White-eyes 159
Whitethroats 153
Wigeon 17, 20
Woodcock 58
Woodlark 96

Wood-owls 77
Woodpeckers 91–94
Woodpigeons 66, 67
Woodshrikes 104
Wood-swallows 104
Wren 138
Wren-babblers 138, 139
Wryneck 91

X

Xenus 55
Xiphirhynchus 133

Y

Yellowhammer 163
Yellownapes 93
Yuhinas *Yuhina* 134

Z

Zoonavena 82
Zoothera 111, 114
Zosterops 159

NOTES

NOTES

NOTES

NOTES